GREGOR MENDEL
Planting the Seeds of Genetics

| Jahr | 1878 | | | | | Beobachtungs-Station | *Brünn* | | | | | | | |
| Monat | Oktober | | | | | Beobachter | *Gregor Mendel* | | | | | | | |

Unmittelbare Ablesung am Barometre						Luftdruck (auf 0° reducirter Barometerstand) in Millimetern				Temperatur		Temperatur des trockenen Thermometers nach Celsius				Temperatur des befeuchteten Thermometers nach Celsius
Thermometer nach Celsius am Barometer	Barometer in Millimetern	Thermometer nach Celsius am Barometer	Barometer in Millimetern	Thermometer nach Celsius am Barometer	Barometer in Millimetern	700 +									38.9	
						6	2	10	Tages-Mittel	Maximum	Minimum	6	2	10	Tages-Mittel	
						40.9	42.3	43.7	42.3	+17.3	+12.4	+14.8	+17.2	+13.5	+15.2	
						42.5	48.0	51.7	47.4	11.3	5.2	10.0	11.3	6.0	9.1	10.6
						52.6	51.0	51.1	51.6	13.3	1.3	3.5	13.0	7.2	7.9	
						51.0	49.6	50.2	50.3	13.4	1.1	3.2	13.3	6.0	7.5	
						50.8	49.6	49.9	50.1	16.0	1.6	2.7	16.0	8.0	8.9	
						50.8	49.2	48.9	49.6	16.8	2.0	4.3	16.2	7.3	9.3	
						47.5	45.3	44.2	45.7	18.6	4.7	6.5	18.3	11.3	12.0	9.1
						43.8	41.3	39.9	41.8	18.6	8.9	11.0	18.3	14.5	14.6	
						40.2	41.1	42.6	41.3	21.2	8.7	10.7	20.9	12.8	14.8	
						42.5	41.1	41.6	41.7	19.6	9.1	11.5	19.3	11.5	14.1	
						44.9	44.9	47.8	45.9	17.9	8.5	10.9	17.8	11.2	13.3	
						49.3	48.3	48.5	48.7	15.8	3.6	5.2	15.5	10.3	10.3	12.4
						49.5	50.0	49.7	49.7	12.1	8.5	10.0	12.1	10.0	10.7	

GREGOR

Abrams, New York
in association with The Field Museum, Chicago

MENDEL

Planting the Seeds of Genetics

Simon Mawer

PROJECT MANAGER: Eric Himmel
EDITOR: Elaine Stainton
DESIGNER: Laura Lindgren
PRODUCTION MANAGER: Kaija Markoe

JACKET/COVER FRONT: Illustration of garden peas
from the Album Benary seed catalogue, 1886.
Reproduced with the kind permission of the
Director and Trustees, Royal Botanic Gardens,
Kew; Mendel's records of weather patterns (detail).
Mendel Museum photograph by Stepan Bartos.
JACKET/COVER BACK: Mendel's microscope. Mendel
Museum photograph by Stepan Bartos. HARDCOVER
CASEWRAP: *Phaseolus coccineus* from *Flora von
Deutschland, Oesterreich, und der Schweiz* Volume Dritter
Band (Volume 3): © The Field Museum/GN90791d.
PAGES 2–3: Mendel's records of weather patterns (July
1883 recording) demonstrate his meticulous
attention to data. INSET: Mendel as abbot of the
Augustinian Abbey of St. Thomas of Brünn (detail).
See p. 84. LEFT: The Abbey of St. Thomas, Brünn

Library of Congress Cataloging-in-Publication Data
Mawer, Simon.
  Gregor Mendel : planting the seeds of genetics / by
Simon Mawer.
    p.   cm.
  Includes bibliographical references and index.
  ISBN 10: 0-8109-5748-5 (hardcover)
  ISBN 13: 978-0-8109-5748-0
  ISBN 10: 0-8109-9262-0 (paperback)
  ISBN 13: 978-0-8109-9262-7
  1. Mendel, Gregor, 1822–1884. 2. Geneticists—
Austria—Biography. [DNLM: 1. Mendel, Gregor,
1822–1884. 2. Genetics—Biography. 3. Genetics—
history. 4. History, 19th Century. 5. History, 20th
Century. WZ 100 M5372m 2006] I. Field Museum of
Natural History. II. Title.
  QH31.M45M39 2006
  576.5092—dc22   2006004366

Printed and bound in Singapore
10 9 8 7 6 5 4 3 2 1

This book is the companion to *Gregor Mendel: Planting the Seeds
of Genetics.* The exhibition and its North American tour were
developed by The Field Museum, Chicago, in partnership with
The Vereinigung zur Förderung der Genomforschung, Vienna,
Austria, and the Mendel Museum, Brno, Czech Republic.

HNA ▋▋▋▋▋
harry n. abrams, inc.
a subsidiary of La Martinière Groupe
Harry N. Abrams, Inc.
115 West 18th Street
New York, NY 10011
www.hnabooks.com

Contents

Ad. nat. pict. in horto Benary.

Chromolith. G. Severeyns, Brux.

ERNST BENARY, ERFURT.

Prologue

Everyone who has taken high school biology knows him: Gregor Mendel, the Father of Genetics. Peas; they know he worked with peas. And they know he was a monk (wrong, but it'll do for the moment), and they know that no one during his lifetime took any notice of his research and it wasn't until after his death, years later, that his work was discovered.

And it's not just high school biology: Mendel's name is everywhere in modern genetics. There's the Mendel University in his hometown of Brno (in the Czech Republic). There's a Mendel Medal awarded by the Czech Academy of Science for discoveries in biology, and another one given for a lifetime's contribution to science by Villanova University in the United States. The German Academy of Natural Scientists has a Mendel Medal too, as does the Genetics Society of Great Britain. Doubtless there are others. His name is attached to the great catalogue of human genetic information, Mendelian Inheritance in Man (available free online, as OMIM). It is even in Webster's and the Oxford English Dictionary, as Mendelism and Mendelian.

And yet, despite all this, we know so little about the man and his work. We know almost everything about Charles Darwin, that other great biologist of the nineteenth century and an almost direct contemporary of Mendel's, because the rich leave traces of themselves—diaries, memoirs, the reminiscences of faithful servants and attentive teachers—and Darwin came from a rich, privileged background. But Mendel was poor, bitterly poor; and the poor leave nothing. Furthermore, Darwin was famous in his own lifetime: he was written about, attacked, applauded, celebrated, derided; he published numerous books and papers and left behind a vast archive of correspondence and notes, as well as a crowd of enthusiastic followers. Mendel published only two major scientific papers, left no archive, gave rise to no school of science, had no pupils. So while Darwin is in the forefront of nineteenth-century science, Mendel stands in the background, an enigmatic figure, the monk whom we can only vaguely glimpse at work in the garden, stout, amiable, short-sighted, and determined. Who was he?

Illustration of peas from the Album Benary seed catalogue, 1876

What exactly did he discover, and, perhaps more important, what did *he think* he had discovered? Above all, how did he become the father of the modern science that concerns us all—genetics?

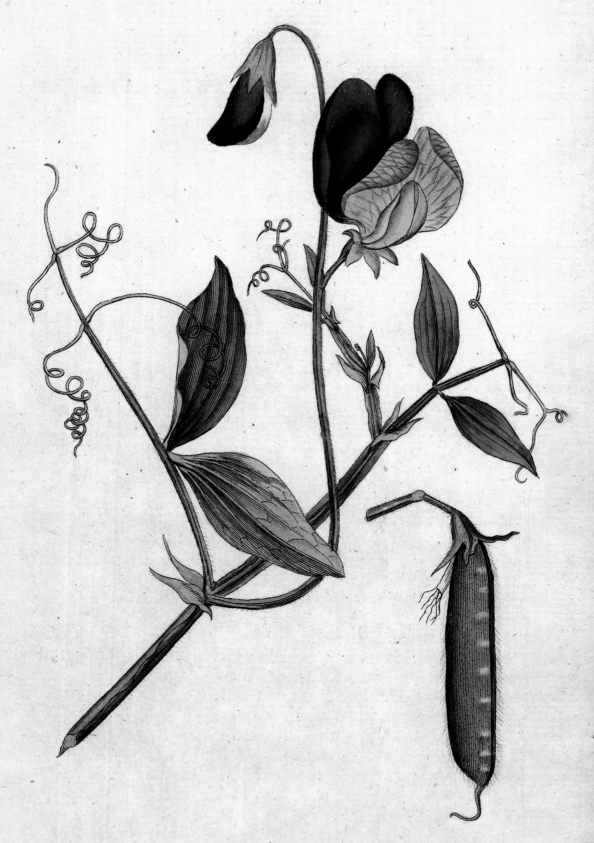

Publish'd by W. Curtis Botanic Garden Lambeth Marsh.

The Lectures, 1865

It would have been cold, that much is certain, for it was February in Central Europe, February 8, 1865. And dark, because it was an evening meeting. Imagine people walking down Johannesgasse (modern Jánská) toward the imposing corner building that is the recently opened *Realschule* in the city of Brünn, Moravia.[1] There might be a horse-drawn carriage or two standing in pools of light thrown down by the gas lamps. Perhaps snow on the cobbles. The hiss of the gas lamps would be audible, and the scraping of horses' hooves. Clouds of condensation hang in the air from the animals' breathing. Men in heavy coats greet each other, stamp their feet against the cold, speculate on the forthcoming lecture, stand aside to let each other go first up the steps and into the building.

In the entrance hall—one imagines he has arrived early and is pacing nervously about as people assemble—is a short, stout figure in clerical black, carrying papers under his arm and, probably, a box of magic-lantern slides. He is Father Gregor Mendel, a friar at the nearby abbey of St. Thomas and, during the daytime, a teacher at this very school. So he is at home here and yet not at home, for this evening he will be lecturing to adults, not children.

Gradually the audience settles in the lecture hall, about forty members of the Brünn Society for Natural Science. There is von Niessl, the secretary of the society; Makowsky, a botanist and geologist who teaches at the Brünn Technical School and is an enthusiastic Darwinian; the physicist Zawadski, who is also a teacher at the *Realschule;* and Joseph Auspitz, the headmaster. They are all worthy and intelligent men, imbued with nineteenth-century progressive views, eager to learn about this strange mechanistic universe that science is unveiling. Sadly, the speaker's superior and mentor, the abbot of St. Thomas's and another founding member of the society, Cyrill Napp, is not here. He is aging and confined to his rooms in the convent.

The audience quiets down expectantly as Karl Theimer—a local pharmacist and amateur botanist with a great interest in plant hybrids—takes the podium. He introduces the speaker, but in truth Father Gregor needs no introduction. Like his abbot, he is a founding member of this society, and everyone knows that over the past eight years

Nobody knows for certain how many pea plants Mendel grew (some accounts say 28,000) or how many peas he sorted (some estimates go as high as 300,000) for his hybridization experiments.

Mendel (front row, second from right) with the *Realschule* faculty in 1864. Like Mendel, many of these colleagues participated in local scientific societies. Some may have been in the audience when Mendel lectured on his pea experiments.

he has been carrying out plant-breeding experiments in the garden at the back of the convent. This evening he is going to give them an account of that work. There is warm applause as the priest rises hesitantly from his chair.

His high-domed forehead shines in the lights as he peers through gold-rimmed spectacles at his audience. He smiles nervously and glances at his notes. "It was my experience with artificial fertilization," he begins, "that led me to the experiments that will here be discussed this evening."[2]

Of course. There are knowing looks among the listeners. Father Mendel is also a member of the local horticultural society and many of its members are familiar with his interest in breeding fuchsias, the flowers that are his favorites.

"Anyone who surveys the work done so far in this kind of breeding will see that among all the experiments carried out, no one has concentrated on the number of different forms that appear among the offspring of hybrids. No one has arranged these forms into their separate generations. No one has counted them." He pauses and nods at the audience. The spectacles catch the light. "No one, in fact, has formulated a generally applicable law that governs the formation and development of hybrids. But doing all this counting and sorting appears to be the only way by which we can finally solve a question whose importance cannot be overestimated."

They stir with appreciation at the word *Bedeutung*, importance. This is what they want to hear. They constitute a society largely made up of enthusiastic amateurs in a city that is less important than Prague, far less than the nearer Vienna. They crave importance.

He smiles wryly. "It requires some courage to undertake a labor of such far-reaching extent." There is laughter. The good friar has a reputation for self-deprecating jokes. Eight years he has been at it, ever since he came back from university in Vienna. They all know. Courage indeed.

Gregor Mendel: Planting the Seeds of Genetics

"So," he says briskly, as though getting down to business. "The first thing was to select the correct plants for such an investigation. From the start I focused on the Leguminosae, the peas and beans, because of their particular floral structure, and finally I settled on the peas because they seemed to have the right characteristics to follow. Some characters don't allow a sharp and certain identification since the difference is of a 'more or less' nature, which is often difficult to define. Such characters could not be used for the experiments; I decided only to work on characters that stand out clearly and definitely in the plants." He continues, warming to his theme, convinced of its importance, convinced that what he has done is the finest piece of experimental work on hybridization that has ever been performed.

ABOVE: Even before his pea experiments, Mendel was known locally for his interest in breeding. Here he holds his favorite plant, the fuchsia, of which he bred many varieties.

Perhaps for these lectures—he is to deliver a second one next month—a magic lantern stood in the center of the aisle and cast its vivid pictures onto the wall. Magic lanterns were popular in those days. Slides were manufactured that could show moving effects—snow falling, people walking, views dissolving from day to night, that kind of thing—so one might imagine that he used lantern slides to show the seven pairs of traits that he finally chose: smooth peas against wrinkled peas; yellow against green seeds; white seed coat with white flowers or gray seed coat with purple flowers; smooth or constricted pods; green or yellow pods; axial or terminal flowers (that is, flowers at the top of the plant, or flowers all along the stem); tall or dwarf plants.

BELOW: When Mendel came to the Abbey of St. Thomas in 1843, Brünn was a dynamically developing industrial town, largely drawing its wealth from the textile industry.

"The flower bud is opened before it is perfectly developed," he explains. "The keel is removed, and each stamen carefully extracted by means of forceps. Once this has been done, the stigma can be dusted over with the pollen taken

from another flower."[3] And then he goes on to detail the initial crossings that he made, one type with another, a total of two hundred and eighty-seven artificial crossings divided among the seven characteristics chosen.

The wise men in the audience nod in approval. This is thoroughness indeed. Some of them have visited the convent and seen the plants, the rows and rows of legumes climbing up their pea sticks, their flowers capped with paper bags to prevent unwanted fertilizations. They have witnessed the care and attention, the obsession.

"Furthermore, in all the experiments *reciprocal* crossings were carried out. By this I mean that each experiment was done twice—once with a particular character in the seed plant, the second time with that same character in the pollen plant."[4] He pauses to let that sink in. "Throughout the whole of the experiment I showed that it is perfectly immaterial whether the dominant character belongs to the seed plant or to the pollen plant; the form of the hybrid remains identical in both cases."

Mendel first announced the findings of his pea experiments to the Brünn Society for Natural Science, which met at the Brünn *Realschule*, a school teaching grades 5 to 10, where Mendel also taught for fourteen years.

Gregor Mendel: Planting the Seeds of Genetics

There is a muttering of interest throughout the audience. Almost casually, almost as a throwaway, the good friar has established as scientific fact what is still only a hypothesis and still doubted by some biologists, that, in terms of inheritance, male and female parents contribute equally to their offspring.

"I also found that in the case of each of the seven crosses, the appearance of the hybrid resembles one of the parental forms exactly. This circumstance is of great importance. I called those characters that are transmitted unchanged in the hybridization, and are therefore themselves shown in the hybrid, *dominant*, and those that become hidden in the process *recessive*."

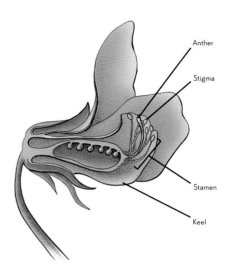

The pea flower.

And so it goes, the stout figure working as methodically through the account of his work as he has through the research itself. He explains how, if you then let the hybrids fertilize themselves, which is what peas normally do, you will get a second generation in which the recessive character reappears. He gives numbers and talks of ratios:

From two hundred and fifty-three hybrids, seven thousand three hundred and twenty-seven seeds were obtained in the second trial year. Among them were five thousand four hundred and seventy-four round or roundish ones and one thousand eight hundred and fifty angular or wrinkled ones. From these a ratio of 2.96:1 is obtained.

Then again, with the different length of stem. Out of one thousand and sixty-four plants, in seven hundred and eighty-seven cases the stem was *long*, and in two hundred and seventy-seven *short*. Hence a ratio of 2.84:1. Incidentally, in this experiment the dwarf plants were carefully lifted and transferred to a special bed, as otherwise they would have perished through being overgrown by their tall relatives. Even in their quite young state they can be easily picked out by their compact growth and thick dark-green foliage.

His figures and ratios continue down the generations and when the lecture finishes the report is still not complete. He will conclude it at the next meeting. The audience breaks up with a scattering of applause, some talking, a nodding of heads. Judgment is suspended.

The second lecture takes place exactly one month later, on the same day of the week—Wednesday—at the same time and with the same group of worthy

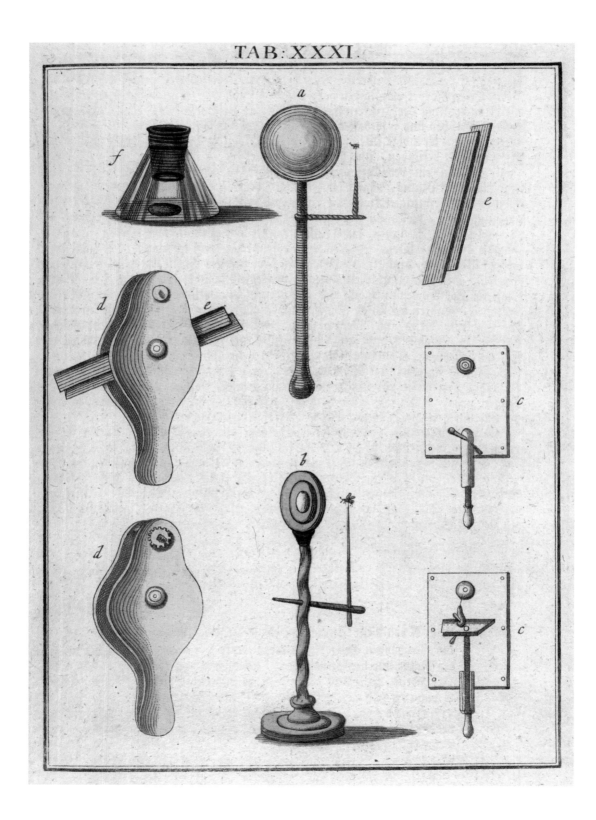

Gregor Mendel: Planting the Seeds of Genetics

scientists and amateur naturalists in the audience. The good friar builds on the results of the first. He lists counts of offspring, explains, argues. There seem to have been thousands and thousands of plants, tens of thousands of peas. It is all most bewildering. He constructs algebraic formulae, hypotheses, rules. He is sure that he has discovered a secret of nature—the rules that govern the hybridization of plants—and he is desperate to communicate his findings. But by now he has, of course, lost them. Figures, patterns, mathematics, assault the audience. They had come to hear about hybrids, monsters created by crossing one species with another—mongrels, mules, or chimeras—and they have got combinatorial mathematics. All very fascinating and impressive—the newspaper report the next morning will agree as much[5]—but all completely confusing. He makes a plea that other members of the society should attempt to repeat his work on other species, but the hope is empty. They applaud at the end; of course they do. It is a splendid piece of work—all those plants, all those figures. "Excellent work," they no doubt assure him. "Excellent." But after a few questions and some discussion—Professor Makowsky speaks enthusiastically about the ideas of the newly published Charles Darwin—Father Mendel, with his patterns and equations, his three-to-one ratios, fades into the background of everyday life.

Today the visitor looks in vain for the hall in which Mendel lectured. The exterior of the building is the same as it was in his time (indeed, it is still a school) and the staircase in the entrance hall has the ponderous weight of an original, but interior walls have been moved, spaces divided, floor plans changed to suit current needs. On the outside wall of the building is a bronze plaque that tells you that the momentous lectures took place here, but inside there is no indication of exactly where the foundations of the science of genetics were laid. And yet there is general agreement that that *is* what happened there in the lecture hall of the *Realschule* of Brno in what is now the Czech Republic, on the evenings of Wednesday, February 8 and Wednesday, March 8, 1865. How did it come about that this retiring, amiable, apparently unexceptional friar delivered a message that would only be understood thirty-five years later and would light a fire that blazes as fiercely today as it ever did? Where did he come from to do such a thing? How did he achieve it?

ABOVE: People had speculated for centuries leading up to Mendel's time about how heredity worked. One common misconception is evidenced in this drawing of a homunculus, a tiny human that some believed was contained in a sperm cell.

OPPOSITE: Early microscopes like these, illustrated in a 1778 German book, opened up an amazing new world beyond what could be seen by the naked eye, but were powerless to probe the workings of heredity.

The Brabant Bellefleur Apple.

Childhood: from Johann to Gregor

The village stands at the very center of Europe, as far from the Baltic as from the Mediterranean, as far from the Atlantic as from the Ural Mountains. In Mendel's day it was called Heinzendorf and it lay in Austrian Silesia, a province of the Austrian Empire. Nowadays the name is Hynčice and it lies in the Czech Republic. Here everything has two names. Even Mendel has two names. Known to the world as Gregor Mendel, he was in fact baptized Johann.

When I first saw Johann Mendel's birthplace it was a summer's day not long after the fall of the Iron Curtain. It seemed a pleasant, sunny place. There were a few dozen farm buildings scattered along a shallow valley, amid orchards of apple and cherry. A stream ran beside the single road; wooded hills rose on either side; a church spire pointed through the trees as though directing the eye upward toward greater things.

Apart from the church, the only building of note stood at the crossroads in the center of the village. At first glance one might have mistaken it for a bus shelter, but there were no prospective passengers waiting for the bus, and a plaque on the roof announced that it was dedicated to the memory of *den hervorragenden Naturforscher u. Klassiker der Botanik, Prëlaten Gregor Joh. Mendel*, the outstanding natural scientist and scholar of botany, Prelate Gregor Johann Mendel. He was born, the inscription specifies, at Heinzendorf, number 58. The building is, in fact, the fire station that Mendel had built for his home village in his later years.

"Mendel?" I asked in the village shop nearby. Fortunately just the name was enough. The shopkeeper closed her till, said something to her only customer, took a key from a drawer, and led the way outside up a muddy path toward the nearest farm building. I waited while she unlocked the door and held it open for me. We nodded and smiled at each other, having no language in common but understanding our respective roles—we were pilgrim and acolyte, united by the fact that here, in this plain, workaday building, on the July 22, 1822,[1] Johann Mendel was born to Rosine Schwirtlich, a gardener's daughter, and Anton Mendel, a peasant farmer.

In the centuries leading up to Mendel's experiments, breeding associations across Europe worked to develop stronger crops and tastier fruits.

There isn't very much to see of the house, just two small rooms and a dusty display of photographs and facsimiles. In the corner of the inner room there is the sole object that might have dated back to Mendel's time—a ceramic stove of the kind that used to be found throughout central Europe, the nineteenth century's version of central heating. Perhaps the small family used to gather round this very stove in the cold of winter, or perhaps not: The house was rebuilt at the turn of the twentieth century—its roof was raised and an extra floor added—and it is hardly the same building that Mendel knew. But, as with any pilgrimage, it is the journey and the witness that matters, not the authenticity of the shrine.

I signed my name in the visitors' book and paid the few crowns entrance fee, then went back outside into the sunshine. The occasional tractor stuttered past along the road. Even fewer cars. Farming is still the dominant way of life in the village and in the summer sunshine there was the illusion that little had changed since Mendel's day, as though his family's descendants might still have been living in the farmhouses hidden among the orchards that he loved so much. But the fact is that in its metamorphosis from Heinzendorf to Hynčice things have changed in a way that it is difficult to comprehend—far more than the mere raising of a roof. To find the peasant boy that was Johann Mendel you have to go back through the history of central Europe. You must cross forty years of Communist rule, go back beyond the Second World War and the short-lived,

OPPOSITE: Mendel shared with many of his countrymen an interest in viticulture and pomiculture. This painting from 1820 shows a grape harvest festival in the hills overlooking Vienna.

BELOW: Mendel's boyhood home as it appears today. Johann Mendel was born July 20, 1822 in Heinzendorf, Moravia, then part of the Austro-Hungarian Empire. The town is now known as Hynčice, Czech Republic.

Gregor Mendel: Planting the Seeds of Genetics

tragic state of democratic Czechoslovakia, back beyond the First World War and into the distant days of the early nineteenth century. Beethoven is still alive. So is Bismarck. James Monroe is President of the fledgling United States of America. The French monarchy has been restored after the upheavals of the Revolution and the Napoleonic Wars. In Great Britain, Princess Alexandrina Victoria is a mere three years old; it will be seventeen more years before she ascends to the throne as Queen Victoria. Closer to Mendel's birthplace, Vienna is a great imperial capital, to rank with London and Paris. From the Hofburg Palace Francis II rules over a vast, troubled, heterogeneous, brilliant, tragic mess of the Austrian Empire[2] that sprawls across central Europe, cutting across languages, nations, and cultures. It includes Italians in the west and Ukrainians and Romanians in the east, it reaches from Poles in the north to Croatians in the south. The center of his realm is filled with Austrians speaking German and Hungarians speaking Magyar, and, subject to these two dominant groups, Czechs and Slovaks. Francis's chief minister is the wily and amoral Metternich, creator of the coalition against Napoleon and architect of the new Europe that grew out of the French emperor's defeat. Francis's family, the Habsburgs, trace their ancestry back to the year 1000. His religion is Catholic. His chief fear is revolution.

The Heinzendorf of those distant days was a tight-knit, traditional, German-speaking, God-fearing place. It was also grindingly poor. Even those farmers who owned some land, as Anton Mendel did, were subject to the *robota*. This was institutionalized forced labor, a kind of serfdom, the last vestige of the old European feudal system. The modern English word *robot* comes, via the Czech author Karel Čapek, from this Slavonic word and perhaps gives some idea of the life that the peasants lived. Under the *robota*, Anton was free to cultivate his own fruit trees and fields for four days in the week, but on the other three he was obliged to work for the local landowner. And when his growing son—a clever, curious child—contemplated his own future, that is what he would have seen ahead of him: a life spent on the farm and in the fields, a life of physical labor and drudgery, as much for the landlord as for himself, a life of brawn, not brain.

The landowner to whom Anton Mendel owed the duty of *robota* was the impressively named Maria Walburga, Gräfin[3] Truchsess von Waldburg-Zeil, an enlightened woman who believed in the education of the people within her jurisdiction. In 1792 she had founded a school for local children at her chateau in the nearby town of Kunin, where the curriculum was based on progressive, practical ideas and included natural history among its subjects. Sadly, the French Revolution had sent shockwaves throughout the Continent and an institution of this kind, one that educated the sons and daughters of the peasants, was bound to come under the suspicion of the authorities. In 1802 the director, Father

Gregor Mendel: Planting the Seeds of Genetics

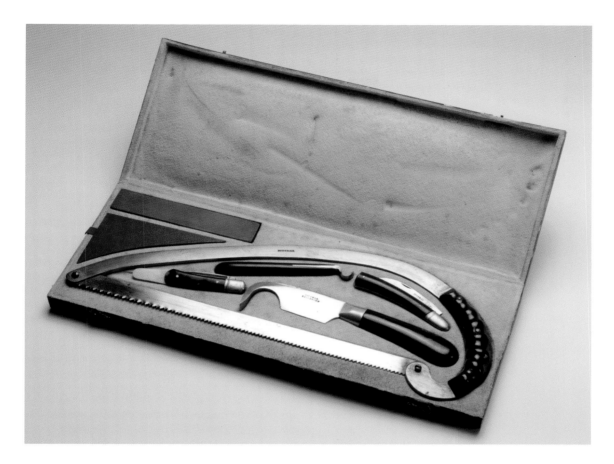

Johann Schreiber, had been removed from his post and the Institute was finally closed in 1814, but its influence lived on in a way that cannot have been suspected at the time: Father Schreiber was demoted and posted as parish priest to the insignificant village of Gross-Petersdorf, where his responsibilities included the neighboring hamlet of Heinzendorf.

The school of Hynčice/Heinzendorf still stands, although it is now a private house. It was a typical single teacher school of the time, where the older children would have assisted in the teaching of the younger ones. It was here that Johann Mendel, his older sister, Veronika, and his younger sister, Theresia, all began their schooling. Naturally the parish priest, an ex-teacher himself, took great interest in what went on in the school.

Biographers of Mendel are left with the merest hints, the softest breaths of coincidence to push them forward: the naturalist Christian Carl André, who later became a founding member of the Brünn Association for Sheep-Breeders, had previously worked as a teacher at the educational institute that was the model for Schreiber's school in Kunin. He knew and applauded the priest's

Mendel's father taught him how to graft fruit trees, sparking a lifelong interest in gardening. These grafting and pruning tools are from Mendel's time at the Abbey of St. Thomas.

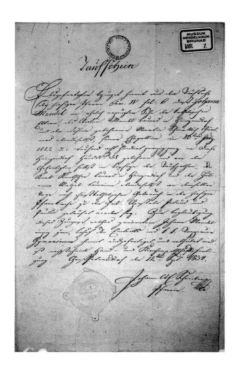

Mendel's baptismal
certificate, 1822

efforts in the teaching of natural science. And Schreiber himself was a founding member of the Pomological (fruit-breeders) Association of Brno and, later, a corresponding member of the Brno Agricultural Society. Out of the latter society would come the splinter group that called itself The Society for Natural Science. Thirty years later it was to a meeting of this society that the priest Gregor Mendel would read his famous paper on inheritance in the garden pea.

Given his interest in horticulture, it is no surprise to find that Father Schreiber kept a small garden beside the schoolhouse in Heinzendorf, where the children were taught gardening and even the rudiments of grafting. Certainly it was here that young Johann Mendel would have learned the facts of pollination and seed germination and the other skills of a plant-breeder.

Father Schreiber was a seminal figure in Johann's life. He baptized the child, oversaw his early education,[4] and on September 12, 1834, copied out the entry from the baptismal register for the fourteen-year-old Johann's admission to the *Gymnasium*[5] at Troppau (modern Opava), twenty-four miles away through the Jeseniky mountains.

Admission to the *Gymnasium* was a turning point in Johann Mendel's life. As the only son of a peasant farmer, he was now committing himself to six years of formal education at Troppau. He would live away from home and therefore not be able to help his father with work on the farm, and would, indeed, be a drain on his family's slender resources. It was his first step in putting a peasant's life behind him and it is not difficult to imagine the arguments and the agony that went into this decision, the persuasive arguments that Father Schreiber brought to bear on Anton, and the discussion between husband and wife over their son's future. But Johann was a talented boy, and in a world where education was the only means of escape, that argument won. He enrolled at the *Gymnasium* on December 15, 1834. As was usual in those days, the headmaster, Ferdinand Schaumann, was a priest. Significantly, he was a friar of the Augustinian Order, a member of the convent of St. Thomas in Brünn. This was Johann Mendel's first direct contact with the place that was to become his home.

So Johann went to live at Troppau, on half-board (bed and a single meal per day) because his family could afford no better, to devote himself to his studies. When the carrier traveled from Heinzendorf to Troppau they sent produce from the farm to augment his meagre rations, but it was not an easy life for the poor boy from Heinzendorf. And it was made worse in the winter of 1838/39,

when disaster struck. His father was out logging in the forests above the village—part of his obligation under the *robota*—when there was an accident. A wagon was being loaded and a trunk rolled free. It is not hard to imagine the sudden panic, the shouts of warning, men slithering and scrabbling in the mud and the snow to hold the great log. They didn't succeed. Anton Mendel was knocked down and lay pinned beneath the trunk. He wasn't killed but ribs were broken and there were internal injuries. They carried him down from the forest a broken man. At Pentecost of 1839 Johann abandoned his studies and returned to help out at home.

A rural family brings a child to be baptized in this scene depicting life in Mendel's native Silesia

There is a family and a personal crisis here, hidden behind the bare words of the record. And there are the first signs of the innate stubbornness that characterized young Johann Mendel throughout his life, for although he did his immediate duty as a son and returned to manage the farm, it was only a temporary measure. Within the year he is back at Troppau, now studying for a private tutor's certificate at the District High School so that he can give private lessons and earn enough money to keep himself. Mendel's first biographer, Hugo Iltis,[6] also mentions some kind of illness at about this time, but there seems little doubt that any illness there may have been was psychological rather than physical. Financial pressures, guilt at the knowledge that he was abandoning his invalid father and the family farm, the stresses of student life—all must have taken their toll on the young man. And yet he stuck to his studies. In 1840 he graduated with success from the school at Troppau and moved on to the university city of Olmütz (Olomouc).

Life in the new city was not easy. As he wrote in his autobiographical sketch in 1850:

He made repeated attempts to offer his services as a private teacher, but all his efforts remained unsuccessful because of lack of friends and recommendations.[7]

Perhaps his new difficulties stemmed from the fact that Olmütz, unlike Troppau, was primarily Czech-speaking and Johann's Czech was rudimentary. Whatever the reason, the struggle proved too much:

The sorrow over these disappointed hopes and the anxious, sad outlook which the future offered him, affected him so powerfully at that time, that he fell sick and was compelled to spend a year with his parents to recover.[8]

Clearly he had suffered some kind of nervous breakdown. Such problems were to dog him throughout his life.

While Johann Mendel was back in his home village that momentous summer of 1841, the matter of Anton's future and the fate of the farm came to a head. One imagines the family meeting round the kitchen table in the dull little cottage, with the two sisters and the young man who had tried to escape to the city. Emotions would have been high. There would have been the dreadful, moral blackmail of family argument, people who know each other too well probing at hidden weaknesses. The father would have talked of duty and loyalty, perhaps even deriding the course that his son had chosen. But Johann remained adamant. He hadn't battled through six years on his own at Troppau in order to surrender now: he was *not* going to take the farm on himself. Here, at the moment when he was at his lowest and most vulnerable, Johann Mendel was prepared to burn any bridges that might have led him back to a life in Heinzendorf.

Fortunately the husband of Johann's older sister, Veronika, showed the interest in the farm that Anton must have sought in his own child. Alois Sturm offered to buy the property. Negotiations were opened with him for the sale and came to completion in a deed signed in August 1841.

Perhaps because there were debts outstanding, the value of the farm was not much, but Anton and Rosine were to be paid a pension in their retirement. And in clause 6 of the sale there is an agreement to make a single payment to the son of the seller of one hundred gulden, along with an annual payment of ten gulden for as long as he continues studying. It is notoriously difficult to give an idea of the value of historical currency, but toward the end of the previous century, in Germany, one source gives a year's expenditure for a student at about

Gregor Mendel: Planting the Seeds of Genetics

five hundred forty gulden, so with the proceeds from the sale of his birthright, Johann Mendel was not gaining any more than a temporary respite from his money worries. What is more interesting is that in the contract there is the first mention of his intention to enter the Church: the money would be paid to Johann "if the latter, as he now designs, should enter the priesthood, or should he in any other way begin to earn an independent livelihood."[9]

With the proceeds from the sale of the farm and with, at last, the promise of some tutorial work, Johann returned to take up his place at the Olmütz Institute for Philosophy. Despite having achieved distinction (*cum eminentia*) in all subjects at the end of the first semester his absence meant that he was forced to repeat the entire first year; but he did exceptionally well, scoring distinction once again in all subjects, except in theoretical philosophy, where he was merely first-class. In the second year exams his success continued, but that didn't solve the problem that was facing him—there was no chance of his being able to support himself through the coming years at Olmütz University despite his younger sister, Theresia's, helping him out with her share of the money from the farm sale. Johann never forgot her kindness in this, and subsequently he supported both her sons through their own university studies; but for the moment her offer still wasn't enough. And now there was no way back, for he had burnt his bridges: the farm had been sold.

Mendel's poverty limited his prospects, but the priesthood offered a stable livelihood and a scholarly life. In 1843, at age twenty-one, he became an Augustinian, taking the name Gregor and entering the Abbey of St. Thomas in Brünn as a novice.

Mendel had an older sister, Veronika (right) and a younger sister, Theresia (left, pictured with her husband, Leopold). Theresia gave up part of her dowry to help her brother return to school.

The door to the future was opened by one of the staff at the Olmütz Institute for Philosophy. Professor Friedrich Franz had recently moved from the southern Moravian city of Brünn (Brno) to replace Andreas Baumgartner[10] as lecturer in Natural Philosophy—that is, physics. He was an able physicist and, in common with so many university teachers in those days, was in holy orders. Although not himself an Augustinian, while in Brünn he had lodged in the Augustinian convent. On July 14, 1843 he wrote to one of the friars:

Gregor Mendel: Planting the Seeds of Genetics

Honored Colleague and very dear Friend!

As a result of your letter of June 12, I have made known to my pupils the Right Reverend Prelate's decision to accept satisfactory candidates at your institution. Up to now, two candidates have given me their names, but I can only recommend one of them. This is Johann Mendel, born at Heinzendorf in Silesia.[11]

Franz goes on to report that during the two-year course, Mendel has had, almost invariably, the best marks, and is a young man of "very solid character." In physics he is "almost first-rate"[12]—an intriguing qualification that perhaps comes from a world where college recommendations were not yet larded with hyperbole. Franz's final observation is that Mendel has some Czech but, given that this is not sufficient, is willing to improve his command of the language during his years of theological study. Could the correspondent inform the abbot about this and ask him what he would like to do about the matter? The abbot replied yes.

Ordination into the Roman Catholic Church has often been an escape from penury for the sons of the poor and Johann Mendel was no exception. He himself admitted as much in his autobiographical sketch, quoted earlier:

... having finished his philosophical studies he felt himself compelled to step into a station of life that would free him from the bitter struggle for existence. His circumstances decided his vocational choice.

So on September 7, 1843, twenty-one-year-old Mendel went to Brünn to be given a medical examination by a Dr. Schwarz in anticipation of his admission as a novice to the Augustinian Order. He was found to be perfectly healthy. On the twenty-seventh he was formally transferred from the diocese of Olmütz to that of Brünn and finally, on October 9, 1843, he was admitted to the convent of the Augustinians as a novice under the name Gregor.

He had escaped.

Pl. 19.

1. Silene quinquevulnera. _ 2. Silene pendula. - 3. Silene purpurea. _ 4. Silene Armeria.
5. Silene vespertina. _ 6. Silene picta. _ 7. Saponaria Vaccaria.

Education

N owadays trams clang and trundle their way through the wide space of Mendolovo Náměstí, Mendel Square in Old Brno. Passengers wait in line at the stops, stamping their feet against the cold. It's a bleak, unlovely place, overlooked on three sides by the tawdry apartment blocks of the Communist era. Only the red roofs of the Augustinian Abbey of St. Thomas, rising over one side of the square, bring some sense of elegance to the scene.

It was here that Mendel came in 1843. He found a remarkable institution, a thriving group of learned and talented men whose lives were committed to the intellectual life of the city. To this day the Abbey of St. Thomas is unusual because the Augustinian Order is not an order of monks like the Benedictines or the Cistercians, but an order of mendicant friars like the Franciscans and the Dominicans. Such religious orders don't have abbeys. Their members live among the community, under a provincial who is responsible to the general in Rome, whereas abbots are more or less independent rulers of their own fiefdom. Furthermore, abbeys frequently own land. So the Augustinians of Brno are, and were, uniquely independent and, in the days when they possessed extensive estates in the countryside around Brünn, powerful. The axis at the center of this power was the diminutive figure of the abbot, Cyrill Franz Napp.

Cyrill Franz Napp

Fifty-one when Mendel joined the order, Napp had been abbot since 1824. He was a true polymath, a small, sharp man with a powerful mind who was equally at ease with the latest developments in science as he was with oriental languages. He was also an efficient manager and an astute politician, defending his own and the abbey's independence against the interference of local politicians and the bishop of Brünn himself. One of his friars, the redoubtable Klácel (see pp. 30, 34, 85), went so far as to describe him as a "secret freethinker." "Freethinker" is a loaded term. In the early eighteenth century it did not necessarily carry with it the strong scent of atheism that it may have now, but it certainly meant someone whose thought was independent and at variance with the established church and state,

Campions (*Silene*). Mendel grew this plant, then part of a genus known as *Lychnis*, in the abbey garden.

Mendel found a supportive scientific community among the abbey's friars, which included philosophers, composers, and scientists. Unlike monks, who lead a cloistered life, friars do public ministry, and the Austrian government of the period required the friars of St. Thomas to teach in public schools.

OPPOSITE: A center of intellectual life, the Abbey of St. Thomas houses a 30,000-volume library dating to the seventeenth century, covering subjects ranging from theology and philosophy to physics and the natural sciences.

and the history of Napp's abbacy confirms this: he was an enthusiastic member of the various scientific societies in the city, a believer in scientific progress, a staunch defender of the less orthodox of his fellow friars, a tenacious fighter against all attempts of church and state to take over the convent, and an advocate of liberal causes in general. But he was also a stickler for form and discipline and he fought from within the establishment. A member of the Estates of Moravia, the regional government, he served on numerous public bodies including the Board of Education for Moravia and Silesia, of which he was director during Mendel's novitiate. He was even, at one stage, deputy Lord-Lieutenant of Moravia.

The group of friars over which Napp held sway was diverse and remarkable for its talents (above). Divided roughly equally between Czech- and German-speakers, they ranged from the revolutionary Klácel to the musical maestro Křižkovsky, from the great Goethe scholar and philosopher Tomáš Bratránek, to the genius of genetics, Gregor Mendel. They were teachers, administrators, university lecturers, writers, mathematicians, philosophers, and natural scientists.

There they are, with the diminutive Napp in the center. This was a community at the very center of life in a vibrant and self-confident city. And they were not afraid of controversy. In 1848, when the people throughout the whole Empire rose up against the repressive and authoritarian rule of Vienna, Tomáš Bratránek joined a group of students marching in protest to the capital. In the same year

Gregor Mendel: Planting the Seeds of Genetics

Education   31

Matouš Klácel was an organizer of the Slav Congress in Prague, a self-appointed assembly that put the aspirations of the Czech people on paper for the first time. So when Mendel joined, five years before those momentous events, he became associated with men who were not only influential in the city and the region, but who were not afraid to speak their minds.

Training and Ordination

Gregor Mendel, in about 1847, the year he was ordained. This is the earliest known photograph of him.

After a period as a novice, Gregor took his final vows of poverty, chastity, and obedience on the day after Christmas, 1846. By then he was studying at the Brünn Philosophical Institute and the Brünn Theological College, covering, among others, such subjects as theology, church history, and biblical archaeology, as well as Greek and Hebrew and other ancient languages of the Middle East. Such was the normal education for the priesthood. But at the Institute the young friar also attended lectures that were of an entirely different significance in his life—those on agricultural science given by Professor Franz Diebl. Like his friend and associate Abbot Napp,[1] Diebl was a member of the Moravia and Silesia Agricultural Society, a student of the latest techniques of plant- and animal-breeding. This is another of those threads that run through the life of Gregor Mendel, linking him to the latest ideas in inheritance. Diebl understood the need to select inherited traits in breeding improved crop plants, and in his major five-volume textbook he describes how artificial pollination may be used to create hybrid varieties. This book would certainly have been used by the students who attended his lectures. In July 1846, at the end of Diebl's course, Mendel achieved distinction in three exams and a certificate of proficiency in pomiculture and viticulture.[2] Besides this official study there was Mendel's own work—in his spare time he spent hours in the herbarium and botanical garden that had been established in the convent some years earlier. The future experimental scientist was beginning to learn his trade.

In the summer of the next year Mendel was ordained in the Dominican church of St. Michael in Brünn. He hadn't yet completed his studies and it was a rushed job, to make up for a shortage of priests: on his birthday, July 22, the day when he actually became twenty-five and was thus qualified to enter the priesthood, he was ordained subdeacon; on August 4, deacon; and on August 6 he was ordained priest. Two months later the new priest was in trouble:

Gregor Mendel: Planting the Seeds of Genetics

To the Very Reverend Prior Baptist Vorthey,

It has come to my knowledge that Father Gregor, having been ordained priest, is attending the lectures in the fourth year of the theological course without wearing a college cap. Father Gregor, although he is now a priest, is still only a student, for which reason and also that he may manifest uniformity *in habitu exteriori* with the other theological students I must ask the Very Reverend Prior to inform Father Gregor that when he attends lectures he must wear a college cap just like the other students.

NAPP.

October 18, 1847

This is just the kind of thing that you might expect from a newly promoted student, but what was to follow the next year was of a different order of things.

1848

1848 was the year of revolutions. Starting in Paris, popular uprisings took place in many of the capital cities of Europe. The demonstrations were against the

The Abbey of St. Thomas, formerly belonging to an order of Cistercian nuns, has been a seat of the Augustinian friars since 1783. Mendel's greenhouse can be seen clearly at the lower right of this rare photo.

ruling, conservative elites, the power of the church, old-fashioned education, old-fashioned ideas, old-fashioned ways. The Austrian Empire was affected more than anywhere else and the students of Brünn, with the redoubtable Professor Diebl at the head, marched through the city singing protest songs—the music of one composed by Mendel's fellow Augustinian Pavel Křižkovský. Abbot Napp voiced his support for the protesters. In Vienna demonstrations had been brutally dispersed by Imperial troops and Napp himself celebrated a requiem mass for the victims at which the pugnacious Klácel preached a sermon calling for the abolition of the *robota* and the introduction of social and political reform. Later there was a counter-demonstration organized by the city's industrialists protesting against the stand taken by the Augustinians. The city was in an uproar.

We have no direct evidence about Gregor Mendel's part in these events but it is not hard to work out. Men were calling for the abolition of the hated *robota*, which his father had suffered, and which he himself would have endured had he not escaped the farm. Furthermore, later in the year six of the Augustinians of Brünn petitioned the provincial Diet for a liberalization of the legal restrictions that governed religious orders. There were five signatories to this protest; Klácel was the first, Gregor Mendel was the sixth. There is some argument about who wrote the petition—one authority has it as Mendel himself—but in all likelihood the author was the first signatory, that well-known progressive, Matouš Klácel. Because of his ideas Klácel had already been banned by the government from teaching and at the time of the petition he was back in Brünn following the violent end to the Slav Congress. It seems likely that he would have been the author of the petition. But certainly his friend, the newly ordained Gregor Mendel, put his name to the document. It was one of the many petitions and protests of that year, and a small part of the big picture, but it shows clearly where Mendel's sympathies lay.

In any event the revolutions of 1848 came to little. Yes, the *robota* was abolished and educational institutions were reformed. Yes, the reactionary Austrian chancellor Metternich was thrown out of power and the emperor abdicated in favor of his nephew, the eighteen-year-old Franz-Joseph. But Franz-Joseph was an autocrat; with his accession a new authoritarian and reactionary regime was established, and censorship was reimposed. Few of the hoped-for reforms remained in place for long.

High School Teacher

Against this uncertain background, Mendel moved on to parish work, which included visiting the sick in the huge hospital just up the hill from the convent.

St. Anne's Hospital is still there on Pekařská (Bückergasse), a ponderous building dating from the end of the eighteenth century. Conditions inside nowadays are very different from those that Mendel found. His was a world before anaesthetics and antiseptics, when death from mere septicaemia was the norm and agony was a constant feature of hospital life.

He couldn't bear it. In a letter from Napp to the bishop we learn that

> he is fitted to study of the sciences, but much less for work as a parish priest, the reason being that he is seized by an unconquerable timidity when he has to visit a sick bed to see anyone ill and in pain.

That much one can perhaps understand, but what is less comprehensible is that "this infirmity has made him dangerously ill and in pain, and that is why I have had to relieve him from service as a parish priest."[3]

There was a weakness there, a fragile streak running through Gregor Mendel's personality. Faced with external pressure—his poverty during the years at Troppau, his experiences in the wards of the hospital, his future attempts to take his teaching examinations—he often suffered some kind of breakdown. Dealing with pupils he seems to have been at ease, but when he was faced with the difficulties of the adult world the young Mendel suffered a number of these crises. Whatever the reason, the outcome of this latest collapse was entirely to his liking: Abbot Napp sent him to the nearby town of Znaim (Znojmo) as a substitute teacher in the local *Gymnasium*.

In teaching, Mendel found his forte. A year later, in a recommendation from the Znaim school, the young friar is described as "an exemplary and safe instructor of youth . . . [who] . . . showed a pure, irreproachable, really priestly way of living . . . and has not used a single word in relation to the moral, religious Church principles or political regulations that was in any way unbecoming or repulsive in a religious sense." Furthermore, "he spent all his spare time in the local reading room except for six visits to the theater, always in the company of a colleague."[4]

The whole report smacks of an intrusive observation of a teacher's life and perhaps is a reflection of the oppressive times in which he lived; but it gives us one of the few contemporary views that we have of him: a shy, introverted young man who is content with his own company. However, Mendel was, it seems, a good teacher, priding himself that he presented his lessons to the pupils in "an easily comprehensible manner."[5]

This panorama of Vienna dates from 1873, about twenty years after Mendel's first visit to the Imperial capital, but it gives a sense of the new world he was entering after a life spent in the countryside and smaller towns.

With the aid of this testimonial Mendel applied for a teacher's certificate. He had to take an examination in physics and natural history and, clearly under-prepared, he failed. The precise reasons for this famous failure are not clear. His physics examiner, Andreas Baumgartner, one-time professor of physics at Vienna University, found the physics paper sound; the natural history examiner disagreed. Whatever the details, Mendel was called to a viva voce examination in Vienna the following August and there occurred one of those dreadful mix-ups that exam candidates have nightmares about. Apparently there was a last-minute

Gregor Mendel: Planting the Seeds of Genetics

change of plan and a second letter was sent from Vienna postponing this interview until the start of the next academic year. Mendel never received this postponement and so turned up in Vienna at the original time.

One can imagine the result. Tight-lipped, the examination board considered his case despite their having postponed the exam. He submitted two written papers—again, in physics and natural history. Again the one was satisfactory, the other a disaster; indeed the natural history paper shows evidence of panic, as though he scribbled down the first thing that came into his head. And following

The pea weevil (*Bruchus pisorum*) was the subject of Mendel's first biological investigation as a member of the Vienna Zoological and Botanical Society. This insect was Mendel's first known connection to the garden pea plant.

that, he had to face the examiners in person. Inevitably he returned to Brünn with his tail between his legs.

This is another fateful moment in Mendel's life. Had he passed his exams and become a fully qualified teacher he might never have made his next step, and had this step not been taken, he might never have turned his attention to the puzzle of inheritance. But he had failed and something needed to be done. So while young Gregor kicked his heels as a substitute teacher at the Brünn Technical School, Abbot Napp made arrangements with Baumgartner for Mendel to attend Vienna University and acquire the formal scientific training that the exam had shown him to be lacking. In November 1851 he traveled to Vienna, found himself lodgings, and began his studies.

University of Vienna

Mendel's Vienna years were instrumental in his development as a scientist. It is here that the one-time farm boy who had shown interest in natural history and knew a bit about plant-breeding and other agricultural techniques began to turn himself into one of the greatest experimental biologists of all time. It seems that his principal talent was in physics. At the newly opened Institute of Experimental Physics, where he attended lectures and practical sessions, the director was the world-famous physicist Christian Doppler (discoverer of the Doppler effect[6]) and, after his death in Mendel's second year of studies, Andreas von Ettingshausen. In 1826 Ettingshausen had already published a textbook entitled *Combinatorial Analysis*. It is precisely this branch of mathematics, the study of combinations, patterns, and probabilities, that Mendel employed in his analysis of hybridization of the garden pea. But he also followed courses in biology (one lecturer was the same man who had failed him in his recent examination) and chemistry. For plant physiology and microscopy his teacher was Franz Unger (see page 45), who, at the very time that Mendel was one of his students, was involved in a bitter public dispute with a conservative Catholic journalist over his own views on evolution.

Using police records—the backlash against the 1848 revolutions was well underway and this was a repressive regime—Mendel's movements among Vienna, Brünn, and Heinzendorf were tracked by Iltis, his first biographer. It appears that he spent vacations at the convent and in October 1852 was back in his native village for the wedding of his younger sister, Theresia. Before that, in July, he had written to a fellow friar about such mundane matters as his laundry and his obligation to attend a forthcoming spiritual retreat that the abbot had arranged. The letter is worth quoting in full, for the insight it gives into Mendel's personality.

Gregor Mendel: Planting the Seeds of Genetics

Dear Anselm,

It is a nuisance that I am once more short of under-linen. No one is in greater need of new under-linen than I am, for of the dozen shirts I brought with me to Vienna as many as 12 are frayed and in holes. Will you please ask Frau Smekal to spend 6 florins in buying linen for 5 shirts, and to get to work upon making them as soon as possible, so that I can at least have one new shirt for the exercises. Would it not be a scandal if the new man I shall become in consequence of the pious exercises were to go about in a frayed shirt? How ashamed I should be if I (Apocalypse:[7] *Stantes amicti stolis albis*—they stood clothed in white raiment) had to parade in torn vesture! The Herr Prelate has already notified me that I am to officiate at the exercises during the last week *hujus*.[8] Since, as you know, the lectures at the university finish on the 20th, and in this matter it would be stupid of me to try to piss in the wind's eye, I have fixed the date of my return for Sunday the 24th, and shall arrive at Brünn toward noon.

Pater Matouš [Klácel] is, I suppose, still in the primeval forests of Trübau. Lucky devil! Leopoldstadt—next week. If tomorrow I win 25,000 fl. as the big prize in the lottery, I shall send a non-committal wire to Frau Smekal. Look her up without fail in the evening! To our speedy and happy (?) meeting![9]

Gregor

It is difficult not to be captivated by the mixture of irony and humor, the crude peasant metaphor combined with the ironical allusion to Holy Scripture. This was a man who had found some kind of contentment, gained some sense of his own place in the world. And he was beginning to find his feet in biology. In 1853 he joined the Vienna Zoological and Botanical Society and read a short entomology paper on a pyralid moth that attacked radishes. The next year he carried out his first biological investigation—into the pea weevil, *Bruchus pisorum* (*Bruchus pisi* to Mendel)—and prepared a second paper that was read to the society in his absence.

This is the first association between Gregor Mendel and the garden pea that was to become so closely associated with his name. Meanwhile his studies came to an end—he didn't take any degree—and he returned to Brünn.

The Realschule

Vienna had clearly given him a thirst for scientific research, but in the meantime he had to justify himself as a newly trained but not yet qualified teacher. Industrialisation was in full flood and Brünn was at the heart of it, expanding as the textile industry boomed. Such a place demands education for its children and new schools were being built at this time. One such was the *Realschule*, a new type of school designed to provide a practical training suited to the burgeoning world of commerce and quite different from the formal and often stultifying education of the *Gymnasium*. The Brünn *Realschule* had opened in 1851 under the headship of Joseph Auspitz and it was here that Gregor Mendel began a career that was to last through his years of research. He gained a reputation as a good teacher and a kindly and good-humored man. "There was no need for him to have recourse to terror as a supplement to instruction," one of his pupils recalled in his old age. This is perhaps a signal virtue given that in those days schoolchildren were beaten, often severely, as a matter of course. He was, it appears, full of enthusiasm for his subjects (he taught physics and natural history) and conveyed this interest to his pupils. Iltis, who had an opportunity in the early nineteen-twenties to interview men who remembered their old teacher, filled many pages of the biography with testimonials to Mendel's popularity and gifts.

The Second Examination

After a successful year at the *Realschule*, Mendel once more applied to the education authorities in Vienna to take the teaching examination. Although we can't be sure exactly what happened on this occasion, the outcome was the same as

Owing to his interest in the natural sciences, Mendel was put in charge of the abbey's collections soon after his arrival. These specimens from the abbey's herbarium date to Mendel's time.

Gregor Mendel: Planting the Seeds of Genetics

This box contains samples of *Pinus silvestris* from the abbey's extensive collections of plants, rocks, and minerals.

the first attempt. Mendel traveled to Vienna in early May and was actually admitted for the exam on May 5, but shortly afterward he returned to Brünn in an unhappy mood and with a distinct reluctance to tell anyone what had occurred. Had he been failed once more, or had he—Iltis suggests this possibility—got into an argument on some technical point with one of the examiners and walked out? We will never know; what is certain is that despite his studies at the university, Mendel never qualified as a teacher and spent the rest of his career officially categorized as a substitute teacher.

Whatever the disappointment he suffered in his second attempt at gaining a teaching certificate, there were other things to occupy him at this time, things that we now see as far more momentous. In the garden at the back of the convent he had been growing plants of many varieties for a number of years—*Fuchsias* were a particular favorite. But now he began to focus on one species in particular, a humble food plant that produces beautiful, butterflylike flowers. Indeed the name of the family from which it comes, the *Papilionaceae*, alludes to that very resemblance: *papillion* is the French for butterfly. In 1856 the young friar planted the first of the garden peas that were to be the basis for his experiments on plant hybridization. In order to understand why he chose this particular project it is necessary to have an idea of where biological thought stood in the 1850s.

The Scientific Landscape

They stare solemnly out at us from black and white photographs, whiskered men (there are no women—you have to wait for the twentieth century for women to begin to take their place in the scientific world) in high collars and elaborate cravats. Their expressions are uniformly stern and uncompromising. They are priests of a new religion, children of the Enlightenment, the liberalizing intellectual movement that had swept through Europe in the eighteenth century and brought with it the French Revolution and the constitution of the United States. Like priests of previous generations, they feel that they are able to peer through the apparent world and view its underlying truth, and they seem confident of what they have found there. Everything is knowable, they seem to be thinking.

But of course the impression given by photographs is notoriously unreliable: in the nineteenth century photography was a new art. Exposure times were long, the chemicals were expensive, and having your photograph taken was an event as significant as having your portrait painted. No time then for candid shots; too much time to hold a fixed smile. So don't be deceived by their stern, unbending expressions. It is important not to misjudge them, and important not to pretend a kind of superiority because we feel we know so much more than they. They were men of passion and intelligence, these early-nineteenth-century scientists, and, one must presume, humor. "Come and see my children," the celibate Mendel would say to visitors—and proudly show them rows and rows of pea plants.

Cell Theory

During Mendel's early years the idea was only just taking firm root that living organisms are machines that you might actually take apart to find out how the separate parts work. Nowadays that is obvious, but in the nineteenth century many still believed that there was something about life that transcended the material world, a "life force" that was peculiar to living beings. Indeed the vitalists held that even the chemicals made by living organisms were somehow engendered by this elusive "force." It wasn't until the synthesis

A detail of the frontispiece of *Wonders of the Invisible World* by Gustav Jäger, 1867. Mendel's books show the range of interests that fed into his research, including evolution, microscopy, and cell studies.

LEFT: German scientist Matthias Schleiden formulated his cell theory in 1839, but believed that the nucleus somehow spontaneously regenerated itself.

RIGHT: German Theodor Schwann worked with Schleiden on his 1839 cell theory. Previous work done by French scientists was largely disregarded because of the dominance of German science at the time.

of urea, the compound that mammals excrete in their urine, in 1828—six years after Mendel's birth—that the first such "organic" compound was made artificially, thus knocking away one of the main supports of the vitalist theory.

A further step was the development of cell theory. Any high school biology student with a half-good light microscope can see cells in plant tissue, and indeed the existence of such cells had been known ever since Hooke first observed a slice of cork under his microscope in 1665. But it is not so easy to see animal cells (as our high school biologist knows, they have no cell walls) and so a general cell theory of living organisms was a long time coming. It wasn't until 1831 that Robert Brown, who had previously discovered Brownian motion (the random movement of minute particles visible under the microscope), also observed the presence of the nucleus in plant cells; and 1839 that Matthias Schleiden and Theodor Schwann formulated the cell theory. In fact Schleiden and Schwann had actually been anticipated by the Frenchmen Henri Dutrochet and François-Vincent Raspail among others, but the antagonism between French and German scientists at this time, and the overall dominance of German science, tended to obscure awkward facts like that. Cell theory is so fundamental to our understanding of biology that it has become a truism, but it bears restating: the fundamental living unit is a cell, and that while some organisms may be made of a single cell, larger organisms are aggregates of many such cells. And it bears emphasising that when this idea was proposed, an idea that we nowadays take for granted, Mendel was already sixteen years old.

Matthias Schleiden in particular got one thing wrong in his thinking: he suggested that the nucleus ("cytoblast") was somehow "crystallizing" out of the intracellular substance and thereby forming a new cell—a kind of spontaneous generation. It took Remak, Virchow, Franz Unger (Mendel's teacher), and others to make clear the principle that all cells come from other, pre-existent cells by some kind of division. By then, in the 1850s, Mendel was just a step away from beginning his work on the garden pea.

With the development of a universal cell theory, other things fell into place. For example, the cellular material was described in some detail by various scientists, and named "protoplasma" by the Czech biologist Jan Evangelista Purkyně. In 1855 Unger proposed that the substance of plant and animal cells is one and the same thing, in other words that all living cells are fundamentally the same. Unger was a true evolutionist, believing that all modern cells have been derived from an original, primal cell. Furthermore he was a clear opponent of the vitalist viewpoint, maintaining that "it is the task of physiology to reduce the phenomena of life to known physical and chemical laws."[1]

Fertilization

It is obvious that before you can understand inheritance you must first understand the process of fertilization in sexual reproduction. It is also obvious that fertilization—one sperm cell fusing with one egg in order to give rise to an offspring—is another of modern biology's axioms. But how was fertilization understood at the time of Mendel? The immediate answer is "incompletely." The significance of Brown's nuclei (the word was his) was not yet clear. There was much speculation but little in the way of empirical evidence.

In 1823 the Italian Giovanni Battista Amici had observed the growth of pollen tubes and later (1830) followed them down the style of the flower to the ovule—one tube to each ovule, he remarked. However, despite seeing bodies within the tubes (which we may presume were the nuclei) Amici was unable to observe penetration into the embryo sac and talked instead of a "prolific humor" being deposited on the surface of the embryo sac and then "imperceptibly absorbed." Certainly he saw the multiplication of cells that resulted from this stimulus and which subsequently led to the formation of the embryo. Later (1841) the German botanist Carl von Nägeli grasped the basic idea of cell division, and, possibly, saw chromosomes for the first time.

All this took place in plants. But ideas were moving forward in understanding animal fertilization. Sperm cells were first seen by that tireless observer Antonie van Leeuwenhoek with his primitive microscope in 1677, but their role was debated. Some thought that they contained the homunculus, the entire future

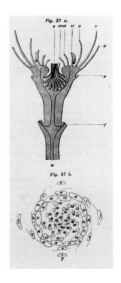

Anatomy and Physiology of Plants by Franz Unger, 1855. Botany professor Franz Unger called Mendel's attention to the newly developing field of cell studies and the statistical relationships between forms of hybrid plants—the key to Mendel's discovery.

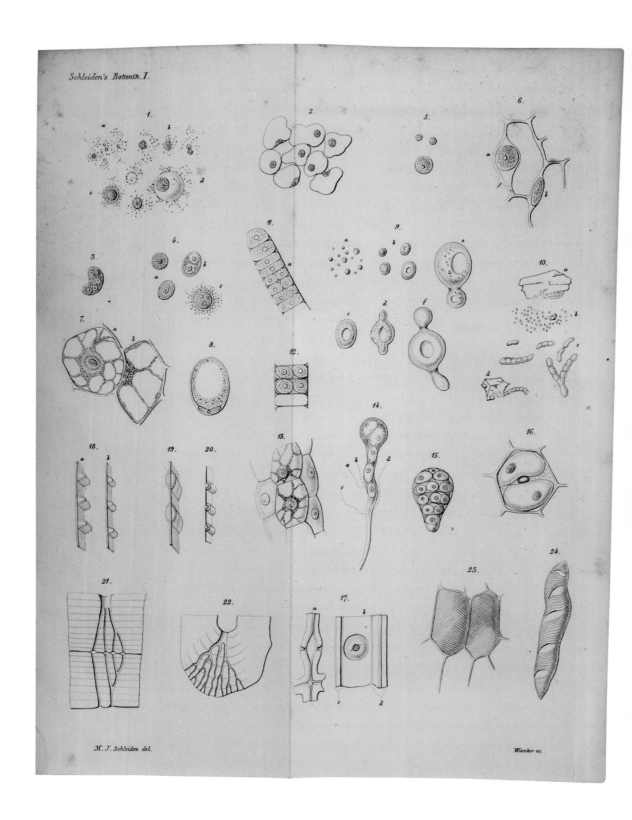

M. J. Schleiden del.   Wienker sc.

being; others thought that a miniature embryo was in the ovum and sperm merely stimulated development. However, in 1825 Jan Evangelista Purkyně observed the nucleus of birds' eggs and named it the "germinal vesicle"; and in 1840 Martin Barry, winner of a Royal Society medal in Great Britain and a co-worker of Schwann's, observed the penetration of a mammalian egg by a spermatozoan, thus supporting Purkyně's speculation about fertilization. Yet there was still debate about how many sperm were necessary for each fertilization, just as there was a similar debate about the number of pollen cells necessary for fertilization in plants.

Purkyně—usually known by the German spelling of his name, Purkinje—was a phenomenon. Coming from a peasant background like Mendel, but with the added disadvantage of being Czech rather than German, he became the first head of the first physiology department and the first laboratory specializing in physiology in the world (both in Breslau, in 1839 and 1842 respectively). His work ranged from the study of human tissues (Purkyně fibers in the heart are named after him) and cell structure to human visual perception. In his speculation about fertilization in 1834, when his wife would have been pregnant with their second son, Karel, Purkyně suggested that "parental characteristics are reduced in the 'germs' to 'pure quality' by a process he defined as 'involution.' He located these germs in the ovum of the female and the semen of the male. The meeting of the parental germs then leads to a new process that he referred to as 'evolution', in which the embryo develops to a form revealing the traits of both parents."[2] This is a priori reasoning to be sure, but it is remarkably percipient.

A significant but elusive piece, Purkyně's work fits into the puzzle that is Gregor Mendel: on four recorded occasions between 1835 and 1850, the great Czech scientist visited the Augustinian convent in Brno. His host was the abbot himself, Cyrill Napp, and he particularly came to see Klácel, but surely such a distinguished visitor would have made his mark on all the members of the community, including the young Gregor Mendel.

Breeding and Hybridization

People have been selecting and breeding plants and animals since the dawn of agriculture. For most of this time, they were working without anything more than a gut feeling about what they were doing, holding back some grain for

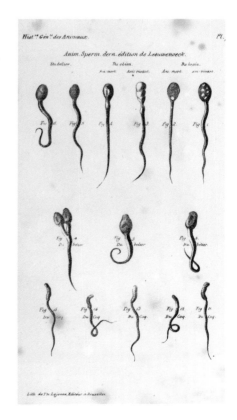

ABOVE: In 1677, long before the cell theorists of Mendel's time, the Dutch scientist Antonie van Leeuwenhoek glimpsed the first spermatozoa under a microscope. The role of these cells in heredity was debated.

OPPOSITE: Cell drawings, from *Outline of Scientific Botany, with a Methodological Introduction as a Guide to the Study of Plants* by Matthias J. Schleiden, 1849

In 1825, Jan Evangelista Purkyně became the first scientist to observe the nucleus of a bird's egg. He later studied a wide range of topics within physiology, including inheritance, and was known to have visited the abbey in Brno during Mendel's tenure there.

planting the next year, keeping the animals they wanted together, and selecting out any weak or otherwise undesirable individuals. Often pure chance helped. For example, an important quality of wheat is that the ears don't easily "shatter," that is, lose their grains. When primitive grain was being harvested it was inevitable that the ears that shattered less readily would be more likely to make their way into the store for grinding and also for next year's planting; by definition the easily shattered ones had already been lost. This led to domestic strains of wheat that would tend to hold on to the grain—a distinct disadvantage in the wild but an advantage for the wheat's survival in the world of agriculture. In preserving such plants our ancestors were carrying out a conscious—if unscientific—artificial selection. A further illustration of chance breeding also comes from wheat. We now know that modern bread wheat, *Triticum aestivum*, is the result of chance hybridizations between various wild grasses, followed by accidental polyploidy.[3] All those first farmers did was to keep seeds from the plants that looked good—they had no idea what they were doing in genetics terms.

Up to the eighteenth century processes like these had led to regional varieties of plants and animals, and a degree of improvement with them, but in general the farmer merely kept the animals in herds or flocks and left the males to get on with it. And get on with it they did, casually and chaotically. But in the eighteenth century—the Enlightenment again—various people began to put selective breeding on a rational footing. And this time there was an external pressure to do so—the Industrial Revolution had begun and with it came a rising, urban population. There were simply more mouths to feed and more people to clothe.

To this end—the improvement of crops—the Englishman Thomas Knight (1759–1838), President of the Royal Horticultural Society, carried out hybridization experiments on a variety of plants using artificial pollination. One of the plants he chose as an ideal experimental subject was the garden pea:

Among these, none appeared so well calculated to answer the purpose as the common pea; not only because I could obtain many varieties of this plant of different forms, sizes, and colors; but also, because the structure of its blossoms, by preventing the ingress of insects and adventitious *farina* (a word for pollen), has rendered its varieties remarkably permanent.

These could almost be Mendel's own words.

In his experiments, Knight clearly demonstrated dominance and recessiveness in the hybrid generation, and the reappearance of recessive characteristics in subsequent generations, but unlike Mendel he did not analyze his results mathematically. Although Mendel was certainly aware of Knight's work (it is cited by Gärtner and was available in German translation), it is not known whether he actually read the Englishman's 1799 paper. However, the influence of Knight is clear, as well as that of other hybridists such as Joseph Kölreuter (1733–1806) and Carl Friedrich von Gärtner (1772–1850), who are specifically acknowledged by Mendel. Like Knight, both these men carried out controlled crossings between differing plants of various species and both found what are essentially "Mendelian" phenomena—that hybrids tend toward uniformity of appearance while their subsequent offspring display a variety of forms, recovering the characteristics of their grandparents. Gärtner even recorded numbers of offspring, discovering in one maize cross what, with hindsight, seems to be a three to one "Mendelian" ratio.

There were others: for example, Goss and Seton, who also worked with peas in England, and Sageret, who bred different strains of melon in France. All reported the phenomenon of dominance and recessiveness and the reappearance of grandparent types in the subsequent generation. None of them actually

PORTRAIT of a TWO-YEAR-OLD RAM of the NEW LEICESTERSHIRE KIND; ..... bred by Mr. Honeyborn of DISHLEY.

In the 1780s, the English livestock-breeder Robert Bakewell pioneered the systematic breeding of sheep and cattle to obtain higher quality wool and fatter beef.

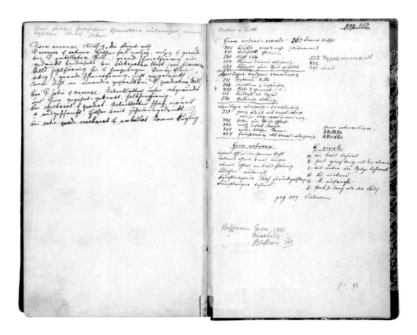

Experiments and Observations of Hybridization in the Plant Kingdom by Carl Friedrich Gärtner, 1849. Mendel studied the work of earlier hybridizers closely. His notes in this copy of Gärtner's book single out the traits of the garden pea that do not vary between generations.

counted numbers or deduced ratios; but one remarkable investigator near to Mendel's home did precisely that.

Johann Dzierzon was a Polish Silesian and a fellow cleric. He had shown that queen bees produce drones with unfertilized eggs, and in 1856 he published work demonstrating how a queen that was a hybrid between Italian and German races of bee produced Italian-type drones and German-type drones in the proportions 1:1. Later in his life Mendel moved on to attempt bee breeding and it is certain that he would have been aware of Dzierzon's work. But whether this gave him the idea of counting ratios in garden peas, we just don't know.

In the animal world, selective breeding had also long been practiced, particularly in horses and dogs, but in livestock pride of place must go to Robert Bakewell. Bakewell (1725–95) was a gentleman farmer in Britain who, in his quest to improve livestock, turned his Leicester farm into what amounted to an experimental breeding station. Up to then sheep had been reared primarily for wool and cattle for haulage—it was Bakewell who set about turning them into high protein food animals, changing their body proportions to raise their useful meat yield and speeding up their time to maturation. Above all, he coupled careful selection of fit specimens with the principle of "breeding in-and-in," that is, inbreeding between immediate relatives in order to create robust and constant blood lines; in post-Mendelian parlance he was creating strains with a high level of homozygosity.

There is a thread, not of wool but of association and influence, that leads from Bakewell and his animal breeding in Leicestershire, all the way to Gregor

Gregor Mendel: Planting the Seeds of Genetics

The Red Must

Mendel's childhood was spent amid farmers and fruit growers.

Mendel in his convent in Brünn. By the end of the eighteenth century Bakewell's example was being followed by breeders in Moravia, a center of the European textile trade. Indeed, in Moravia there was a whole collection of agricultural societies, clustered around the parent body, the Moravian and Silesian Agricultural Society. In 1814 a Sheep-Breeders' section was established, to debate and apply the latest ideas that originated from Bakewell's Leicester farm. In 1825 Abbot Cyrill Napp became a member (later president) of the Agricultural Society and an active participant in the sheep-breeders' meetings. His interest was both scientific and commercial—more than 50 percent of the convent's income was though the wool trade—but above all, it was enthusiastic. They argued round and round, these worthy men. They were farmers, breeders, mill owners, and textile magnates. They often used language that was orotund and imprecise, clouding ignorance and misunderstanding in a fog of obfuscation, but occasional clarity shone through. It was Abbot Napp himself who declared— his very words are reported in the minutes of one meeting in 1837—"the question for discussion should not be the theory and process of breeding, but *what is inherited and how*?"[4]

Six years after he had set this agenda, it was Cyrill Napp himself who welcomed Gregor Mendel into the Augustinian community as a young novice. Napp wasn't to know it then, but within a decade it would be Mendel who would take up the abbot's challenge.

XVII, 3. 106. *Leguminosae.*

453. *Pisum sativum L.* Brech-Erbse.

The Research Program

Any biology textbook will give it to you, the whole story of what Mendel found—but a textbook will give you the twentieth- or twenty-first-century idea. Much of it will be misleading; some of it will be plain wrong. Without a doubt, Mendel himself is the best guide to his research program—his paper of 1866 is a perfect, lucid account of his work—but it is still possible to summarize what he did.

In 1854 (probably—it is not easy to be certain) Mendel obtained thirty-four different varieties of pea seeds from local seedsmen, and then spent two years growing and breeding specimens to see what characteristics would be appropriate to use. He was looking for clear-cut, "either-or" differences, i.e., clear examples of what we would nowadays call *discontinuous* variation. Mendel used the phrase "constant differentiating characters." In doing this he was remarkably discerning. Almost all other hybridists of the period worked with less precisely defined characteristics that showed continuous variation—for example, "size," or, even worse, "appearance." The second key choice was to abandon merely "qualitative" observations—i.e., the use of imprecise words such as "many" or "few"—and to go for quantitative observation—precise counting. Mendel had attended Ettingshausen's lectures on combinatorial analysis. He understood the mathematics of probability and outcome.

Of the thirty-four varieties that he started with, he finally narrowed his choice down to twenty-two types from which he actually chose seven for experimental work: 1) smooth vs. wrinkled peas; 2) yellow cotyledons with yellow peas vs. green cotyledons with green peas; 3) white seed coats with white flowers vs. gray seed coats with purple flowers; 4) smooth vs. constricted pods; 5) green vs. yellow pods; 6) axial vs. terminal flowers; and 7) tall vs. dwarf plant. However, all twenty-two types were bred throughout the whole of the experimental program to show that, left to their own devices, they all bred consistently true.

An obvious question is, why did he choose the garden pea? It does seem that *Pisum sativum* is especially suitable for what he wanted. For example, it is naturally self-pollinating. That is, the pollen of a flower is transferred to the stigma of the *same* flower before the flower bud opens—thus ensuring that the seeds formed have the one plant as both male and

The garden pea (*Pisum sativum*)

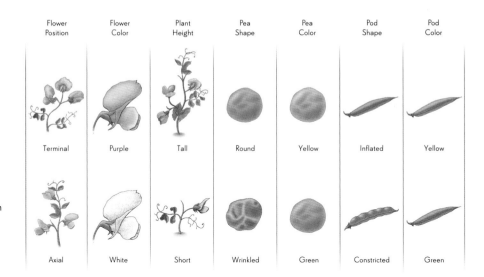

Flower Position	Flower Color	Plant Height	Pea Shape	Pea Color	Pod Shape	Pod Color
Terminal	Purple	Tall	Round	Yellow	Inflated	Yellow
Axial	White	Short	Wrinkled	Green	Constricted	Green

Mendel chose seven pairs of contrasting traits to observe in his experiments: smooth vs. wrinkled peas; yellow vs. green peas; white seed coats with white flowers vs. gray seed coats with purple flowers; smooth vs. constricted pods; green vs. yellow pods; axial vs. terminal flowers; and tall vs. dwarf plants.

To cross-pollinate pea plants, Mendel removed the anthers from within a flower before they made pollen. He then carefully brushed the flower's stigma with pollen from another plant. The cross went both ways, wherein each plant was fertilized with pollen from the other.

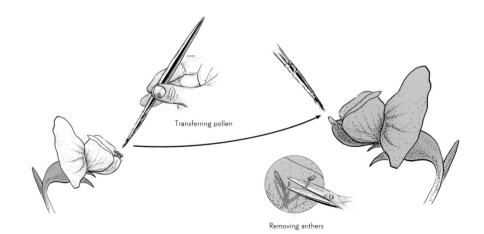

Transferring pollen

Removing anthers

female parent. This means that pea varieties are naturally "pure." And as the flowers are relatively large, it is easy to open the immature flower buds, remove the anthers before they are ripe (there are nine pollen-producing anthers in total, arranged around the end of a tube that sheathes the ovary), and so prevent the flower from being able to fertilize itself. Once the anthers have been removed, the flower is ready for pollination using pollen from another plant altogether—the plant of the experimenter's own choice. Pollen from this plant would be transferred with a camel-hair paint brush to the stigma of the emasculated flower. Mendel, like many an artist, brushed his way to discovery; and like many an artist, he was painstaking in his application. Each cross went both ways, pollen of one type being transferred to the stigma of a flower of the other type, and the reverse—that is, pollen of the second type of plant also being transferred to the stigma of the first. This is known as reciprocal crossing. This alone was a demonstration of major importance, confirming empirically what until then had only been suspected theoretically (and was denied by some): that each sex contributes equally to the offspring.

Once the pollen had been transferred, the flowers were enclosed in calico or paper bags to prevent any further pollination; and then there was nothing to do but look after the plants and wait for the fruit (the pods) to mature. The pea is an annual plant, growing from seed in the spring, flowering and setting fruit, then withering and dying. It is all over within about twelve to eighteen weeks and the dried seeds (peas) are all that remain of the plant through the autumn and winter.

List of seeds ordered by Mendel from the Ernst Benary seed firm. In 1856, Mendel launched his ambitious series of experiments with the garden pea.

So for Mendel one generation equalled one year. Once he had collected the seeds, he would dry them and store them until the next spring.

His organization and labeling were meticulous. He planted the seeds in beds in the convent garden, but also reared some plants in pots. In particular, some were confined to a greenhouse as a control population (Mendel's use of the word *control* in this scientific context was almost certainly an all-time first). This control population was to ensure that there were no chance cross-fertilizations among plants grown in the open, perhaps through the action of the pea weevil, which attacks pea flowers. No difference was ever found between the greenhouse populations and the open-bed populations. His method was secure.

The greenhouse that Mendel used was built for him by Abbot Napp at the start of his work (showing how seriously Napp took the investigation). Sadly, it was destroyed in the great storm of 1870, but when he himself was abbot, Mendel apparently rebuilt the greenhouse and it still stood at the beginning of the twentieth century. Today, however, nothing but the foundations remain on the lawn.

1856. The Opening

By the spring of 1856 the preparatory work was over and the suitable varieties had been chosen. The main experiment could begin. It started on a small enough scale, with Mendel planting each of the differing varieties, and then, once the flower buds had appeared, making the initial cross-pollinations. There must

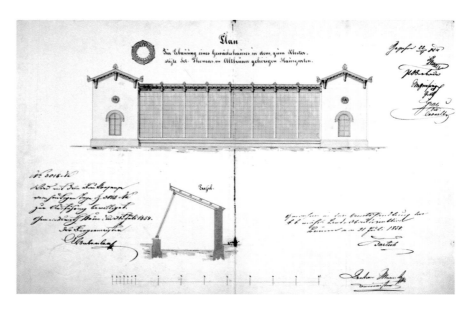

Abbot Napp had this greenhouse specially built for Mendel's experiments in 1854, demonstrating how seriously he took the young friar's work.

have been a feeling of anticipation, a snatch of excitement as he began the task, choosing the flower buds, gently removing the keel, peering through a hand lens at the delicate parts inside, and carefully plucking off the immature anthers with a pair of fine forceps, and then quickly dusting the remaining stigma with pollen from the chosen male parent. An intricate, repetitive procedure:

EXPERIMENT 1 smooth peas crossed with "angular" peas (these days we tend to say "wrinkled"): 60 fertilizations on 15 plants.

EXPERIMENT 2 yellow cotyledons crossed with green cotyledons: 58 fertilizations on 10 plants.

EXPERIMENT 3 white seed coat crossed with gray seed coat: 35 fertilizations on 10 plants.

EXPERIMENT 4 smooth pods crossed with constricted pods: 40 fertilizations on 10 plants.

EXPERIMENT 5 green pods crossed with yellow pods: 23 fertilizations on 5 plants.

EXPERIMENT 6 axial flowers crossed with terminal flowers: 34 fertilizations on 10 plants.

EXPERIMENT 7 tall plant crossed with dwarf plant: 37 fertilizations on 10 plants.

Assuming his own stated average of six to nine mature peas per pod, even that gives a total of over two thousand seeds to harvest and to plant the next year. But that is just the beginning.

1857. The Hybrids (F_1)[1]

All the peas from the first crossings were planted out and grown to maturity. *In each case they appeared exactly like one of their parents* (and therefore quite unlike the other). For example, in Experiment 7, all the hybrid offspring were as tall as their tall parents (actually, Mendel meticulously noted, on average they were very slightly taller). He called the characteristics that appeared in the hybrids *dominant*, and the ones that disappeared, *recessive*, which is, of course, the origin of these universal terms in genetics. In order to talk about these plants he needed a simple notation, and this is what he chose: the dominant plants he started with would be noted with a capital letter, or **A**, and the recessives with a lower case letter, or **a**. He never used letters that were related to the characteristic in question (for example *T* for tall and *t* for dwarf, as is found in modern textbooks) but simply took the first letters of the alphabet in order. The hybrids that he had created by artificially crossing one with the other would be **Aa**. If he was considering two characters together at the same time he would simply use the next letter of the alphabet. Thus **AB** and **ab**, or for a double hybrid **AaBb**. It is worth noting here that these letters referred to the appearance of the plants, not, as a geneticist would nowadays use them, to the genetic makeup of the plants. Mendel knew nothing at all about genetic makeup. Of the seven characteristics he was using, the following are the dominant ones: 1) round seed; 2) yellow cotyledon with yellow seed; 3) gray seed coat with purple flower; 4) smooth pod; 5) green pod; 6) axial flowers; and 7) tall plant.

In fact seed shape and color could be seen in the pods of the parent plants, and so he didn't need to wait for the following year to see these features; and because they were features of the seeds, rather than the fully grown plants, they also gave him large numbers. These two facts explain why he tended to concentrate on these features in the later stages of the experiment and why they always appear in larger numbers in his results.

Initial cross

First generation

Second generation

Third generation

One of Mendel's most important discoveries was what he termed "dominant" and "recessive" traits. When Mendel first bred a green pea plant with a yellow pea plant, the dominant trait masked the recessive trait, creating all yellow peas in the first generation of offspring. The recessive trait reveals its continued presence in the green peas of the second generation.

1858. The First Generation from the Hybrids (F_2)

After growing the F_1 plants, he did nothing beyond looking after them. No more artificial cross-pollination at this stage. He merely allowed the flowers of the hybrid generation to self-pollinate as they would do naturally. Quite naturally, therefore, they set seed and fruit and once again he collected the seeds (and, of course, noted immediately the seed shape and color of the next generation—see pp. 56–57), dried them, labeled them, put them away for planting the next spring. Meanwhile, life went on. Daily he walked over the hill from Altbrünn to the *Realschule* on the other side of the city center, trudging through the rain of autumn and the snow of winter, a short, bespectacled figure in black coat and top boots. And back in the greenhouse, in carefully labeled boxes, the dried peas waited. Thousands of them. When the snow melted in spring and the temperature rose, he could prepare the beds and the pots, and at the right moment begin planting. Then he would wait while the plants grew, at first mere shoots out of the earth, and then quickly straggling, random things, grasping anything they could to hoist themselves upward, like children learning to walk, pulling themselves off the ground and clinging to something stable in order to stand. Mendel and the gardener would provide the crutches, going round the beds inserting pea sticks and attaching strings. And counting. And in this second generation the hidden characters reappeared. Of course it is what he had expected: he had already counted the seed shape and colors. But one can imagine the steady satisfaction as the pattern revealed itself, the recessive characters reasserting themselves and in the numbers that his understanding of combinatorial analysis had probably already suggested to him:

These are his results from the F_2 generation:

Character	Total	Dominant	Recessive	Ratio
Seeds				
Shape of seed	7324	5474	1850	2.96:1
Color of seed	8023	6022	2001	3.01:1
Whole Plants				
Color of seed coat	929	705	224	3.15:1
Shape of pod	1181	882	299	2.95:1
Color of pod	580	428	152	2.82:1
Flower position	858	651	207	3.14:1
Height of plant	1064	787	277	2.84:1

He would, of course, have started explaining what was happening, both to himself and to others. No doubt Abbot Napp was called to witness and understand. And Klácel and Bratránek, and Lindenthal, the brother who stands next to Mendel in the group photo (p. 11) and who appears to be looking at the fuchsia

Gregor Mendel: Planting the Seeds of Genetics

that Mendel is holding. Lindenthal would certainly have been in on the work because, according to Iltis, he actually assisted Mendel, and in 1859 was proposed for membership in the Natural Science Society by Mendel himself. But in his famous paper, Mendel leaves the explanations to the second half, and merely moves on to the next generation.

1859. The Second Generation from the Hybrids (F_3)

Again, the collection of the seeds for drying. Again, the counting of the seed shapes and colors, and with that count, a revelation of what was to come next year. Again the wait over winter, during which the ordinary life of a teacher and a friar went on. Mass in the convent chapel in the morning. The trudge over the hill to and from the school, the round of services in the great Gothic church of the Blessed Virgin on Sundays. And marking schoolwork in the evening, going over the shoddy writing, the half-answers by incomplete minds, the banalities, and the drudgery. And all the time the peas waited in their boxes, dry, lifeless, but full of potency.

In this generation it was shown that all the recessive types from the previous year (Mendel called them type *a*) had produced nothing but recessive offspring, i.e., they were pure. Of the hybrid plants that had all exhibited the dominant type, one third were now shown to have been pure (he called them **A**), while the remaining two-thirds again produced dominant and recessive offspring in a three-to-one ratio, showing that they were hybrids (he called these hybrids **Aa**). With experiments 1 and 2, this could be found merely by looking at the seeds that the F_2 plants produced, but to demonstrate the same effect in the other experiments whole plants had to be grown. Mendel therefore selected one hundred of each of the these dominant F_2 plants (experiments 3, 4, 5, 6, and 7) and planted ten seeds from each plant. This alone yields *one thousand* plants for observing the next year! Furthermore, he kept the lines going through until the end of the whole program. The number of plants grown begins to climb and by now Mendel would have taken over much of the convent garden. Rows and rows of peas would have been inching their way upward in the spring. The work of tying them up, watering, weeding, then harvesting, sorting, and drying the peas, would have become a major call on the gardener's time.

The Subsequent Generations

From this point on it appears that Mendel's investigation pursued three arguments. The degree to which these were independent of each other is debatable:

The second generation of Mendel's pea hybrids each had one green pea gene and one yellow pea gene, as shown in the parent peas pictured here. When these parents pass their genes on to the next generation, they are likely to produce one green pea for every three yellow peas. This ratio is illustrated here in a Punnett Square, invented by R. C. Punnett in the early 1900s.

1) The Continued Inbreeding of the Dominants (**Aa** x **Aa** or **AA** x **AA**)

He continued to allow some dominant plants from each experiment to self-pollinate and produce young. This was kept up for generation after generation, with some lines continuing right down to 1863, when the work effectively stopped. Each generation of dominants did what was expected of it, i.e., one-third produced only more dominants; two-thirds produced dominants and recessives, thus showing that they were hybrids (**Aa**). Experiments 1 and 2 were carried through six generations; 3 and 7 through five; and 4, 5, and 6 through four. From this Mendel deduced a rule, that if you hybridize two varieties and then allow them to breed down the generations (assuming four offspring per plant) you get the following pattern:

		Numbers			Ratios		
Generation	Generation No.	A	Aa	a	A	Aa	a
F_2	1	1	2	1	1	2	1
F_3	2	6	4	6	3	2	3
F_4	3	28	8	28	7	2	7
F_5	4	120	16	120	15	2	15
F_6	5	496	32	496	31	2	31
n					$2^n - 1$	2	$2^n - 1$

One may see from this that although the number of **Aa** hybrids doubles each time, as a proportion of the total number of plant hybrids actually decline—in terms of the *ratios* the hybrids remain constant at a value of 2.

This observation, which is mathematically correct, has almost no importance in biological terms. Yet it was seen as significant by Mendel and it probably accounts for the major misunderstanding of the significance of Mendelism in the early part of the twentieth century. Mendel himself thought it mattered because at the outset he was concerned with the process of hybridization, and this shows that, left to their own devices, self-pollinating hybrids will tend to revert to parental type, although the hybrid form will never actually disappear. However, even though he himself didn't fully appreciate it, by this time his investigation had ceased to be about hybridism per se and had become something far more important: the first true genetic investigation in the history of science.

2) Two or More Different Characters Inherited Together

From 1859 Mendel also began to consider *combinations* of two or more characters together on the same plant. It now becomes impossible to calculate with any accuracy the numbers of plants involved, partly because it may be that he

Gregor Mendel: Planting the Seeds of Genetics

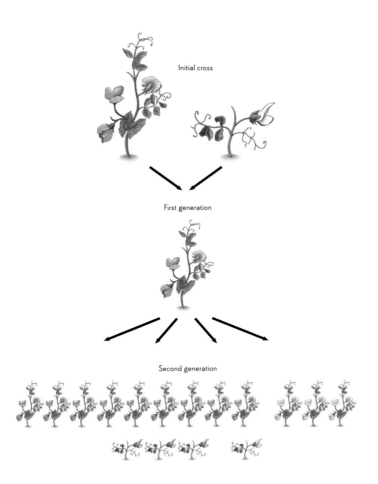

Initial cross

First generation

Second generation

Mendel also crossed plants that differed in more than one trait and found that the second hybrid generation showed every possible combination of traits in a 9:3:3:1 ratio. Mendel realized that the factors governing traits traveled completely independent of each other—seed color was not linked to seed shape, height was not linked to blossom color—when they were passed along to the offspring. This is now known as the Law of Independent Assortment.

was already doing these combinations in his original experiments and merely separated out the results of them in his paper. For example, he had set up a line with seeds that were both *yellow* and *round* (**AB**), and another with seeds that were both *green* and *angular* in shape (**ab**). In another experiment he had plants with *dwarf* stems, *terminal* flowers that were *white* in color, and pods that were *smooth*; these he crossed with plants that were *tall*, with *axial*, *purple* flowers and *constricted* pods. Here, obviously, there are *four* different genetic characters together on each parent plant. Then he crossed one line into the other, and what he demonstrated was that each pair of characters operated independently from the other pair—i.e., produced its own 3:1 ratio and was unaffected by the other characters in the plant. That is, the two characters are inherited independently of each other. Thus the hybrid with round seed with yellow seed (**AaBb**) when self-fertilized produced: 315 round and yellow, 101 wrinkled and yellow, 108 round and green, 32 wrinkled and green.

Two 3:1 ratios put together—(3:1) x (3:1)—gives a 9:3:3:1 ratio, something that everyone who studies Mendelian genetics comes across sooner or later, and surely one that Mendel himself must have understood (as you can see, the example given above is a good fit to a 9:3:3:1). However, nowhere in his writing does he actually specify this ratio, although he does list all the possible combinations of characters and the number that would appear. What he does do, at this point in his paper, is produce one of his rules: "the offspring of the hybrids in which several essentially different characters are combined exhibit the terms of a series of combinations, in which the developmental series for each pair of differentiating characters are united."[2] It is demonstrated at the same time that "the relation of each pair of different characters in hybrid union is independent of the other differences in the two original parental stocks." This, of course, is the idea of "independent assortment": One gene pair is inherited independently of another gene pair. It is also—but discussion of this comes later—not necessarily true.

In this section of the work he goes on to consider three, four, and more of his characters together on one plant. Reading the paper you feel that he is unstoppable:

> If n represent the number of the differentiating characters in the two original stocks, 3^n gives the number of terms of the combination series, 4^n the number of individuals that belong to the series, and 2^n the number of unions that remain constant. The series therefore contains, if the original stocks differ in four characters, $3^4 = 81$ classes, $4^4 = 256$ individuals, and $2^4 = 16$ constant forms: or, which is the same, among each 256 offspring of the hybrids are 81 different combinations, 16 of which are constant. All constant combinations that in peas are possible by the combination of the said 7 differentiating characters were actually obtained by repeated crossing. Their number is given by $2^7 = 128$.

Despite the complexity it is worth pausing to consider this. What he is saying here is that if you take two plants that differ from each other in *each* of the seven original character pairs and you cross-pollinate them, yes, the hybrids will all look the same (they will show each of the seven dominant characters); but when the hybrids self-pollinate to produce the next generation (the F_2) you will obtain *one hundred and twenty-eight different combinations* of the characters. Furthermore, he claims that he *actually did this*—which gives weight to the suspicion that he was using combinations of characters from the very start of his breeding program. But the real significance of this finding was missed by everyone, even when Mendel's work was "rediscovered" in 1900: this is precisely the inherited variation that Darwin needed to make his theory of natural selection work. It was there, in the published literature, at exactly the time that Darwin was pro-

Gregor Mendel: Planting the Seeds of Genetics

posing some absurd system of gemmules floating around in the blood and being picked up by the sex organs (see pp. 91–93).[3]

3) The Backcross to the Recessive

The third direction Mendel took was to investigate the composition of the egg and pollen cells of the hybrids. This is, essentially, Mendel's *explanation* of his observations. It is where he comes so tantalizingly close to inferring the existence of genes. His reasoning is that if the double hybrids (**AaBb**) when self-pollinated (**AaBb** x **AaBb**) can make all the possible combinations of offspring, it must be because each plant has produced pollen cells or egg cells of all the possible types and in equal numbers: i.e., 25% **AB** + 25% **Ab** + 25% **aB** + 25% **ab**

He tested this by crossing such a hybrid with a double recessive plant **aabb**, and yielded the ratio 1:1:1:1, which is now known as the "test cross."

		Double Hybrid (AaBb)			
Double Recessive (aabb)	gametes	AB	Ab	aB	ab
	ab	AaBb	Aabb	aaBb	aabb
	numbers:	47	38	40	41

The importance of this part of the program in the understanding of what Mendel thought he had discovered will be considered later.

Meanwhile, the work went on. Numbers of plants? Impossible to estimate with any certainty. Fisher's paper of 1936[4] suggests over five thousand plants for 1859, and over six thousand for 1860. It has become bewildering. Row upon row of peas grow in the garden behind the monastery. The greenhouse is working full time. Obsession? Reading his paper it seems that it must have been. Throughout each spring and summer the man spent hours and hours tending his plants, pollinating, scoring, labeling, harvesting, drying, putting seeds away for the next year, peering at the world through his gold-rimmed spectacles, puzzling and pondering, counting, and tallying, explaining to anyone who would listen what was going on (many did; none really grasped the significance). Visitors were in the presence of a man inspired—a Beethoven or a Goethe—and all they saw was a dumpy little friar with a sense of irony introducing them to his "children." Sublimation? Perhaps. And yet there is no doubt that what he did was, for its time, absolutely and completely remarkable and unique. To give you an example of the "next best," Charles Darwin himself reports on a breeding experiment that he did with two distinct forms of the snapdragon. One, the "peloric," has radially symmetrical flowers, that is, flowers with many lines of symmetry, like a wheel; the other is the normal form with

Snapdragon (*Antirrhi-num majus*)

bilateral symmetry, having only a single line of symmetry, like a human being. Here is Darwin's report of the experiment, from his book *The Variation of Animals and Plants under Domestication* of 1868. It is worth bearing in mind that he was doing this experiment at exactly the same time that Mendel was at work:

Now I crossed the peloric snapdragon (*Antirrhinum majus*), described in the last chapter, with pollen of the common form; and the latter, reciprocally, with peloric pollen. I thus raised two great beds of seedlings, and not one was peloric. The crossed plants, which perfectly resembled the common snapdragon, were allowed to sow themselves, and out of a hundred and twenty-seven seedlings, eighty-eight proved to be common snapdragons, two were in an intermediate condition between the peloric and normal state, and thirty-seven were perfectly peloric, having reverted to the structure of their one grandparent.[5]

There you have it: the great man has made an artificial cross and produced hybrids that show complete dominance. Then he allows the hybrids to self-pollinate, gathers the seeds and sows them, and finds that the peloric form reappears in the F_2. It is pure Mendel. He has even got an approximation to a 3:1 ratio (2.38:1 in fact) with a distinctly small sample. And as for the two intermediate forms that Darwin found, well Mendel had to deal with the occasional problem of that kind, but just skirted round it. So having got this far, what does Darwin do? The answer, I am afraid, is nothing. That is not the measure of Darwin: it is the measure of Mendel.

Further Work

Of course, Mendel didn't finish there, but that is the essence of his work on peas. He did try similar experiments with other species, in particular *Phaseolus*, the bean. In these experiments (they are mentioned at the end of his paper) he found confirmation of the *Pisum* rules with a number of characters, and, when investigating flower color, a curious new possibility: When he crossed the white-flowered form with a purple-red-flowered form, all the hybrids were slightly less intensely colored than the purple-red parent. However, on self-pollinating these, the resulting F_2 showed a wide range of color in the flowers, from deep purple-red to pale violet. Unperturbed by this, Mendel suggested that "even these enigmatical results might probably be explained by the law governing

Pisum if we might assume that the color of the flowers and seeds of *Ph. multi-florus* is a combination of two or more entirely independent colors, which individually act like any other constant character in the plant." He goes on to elaborate the idea using characters A_1, A_2, A_3, and a_1, a_2, a_3, etc., all joining together to achieve the one effect of flower color. In fact what he has done is give a good description of what we would nowadays call polygenic inheritance.

What Did Mendel Actually Discover?

Perhaps one might begin with some of what he did *not* do.

He did not discover the "gene." He did not even discover something—a "factor"?—that later would be called the "gene."

He did not formulate, in explicit terms, either "Mendel's" first law (segregation) or "Mendel's" second law (independent assortment). In fact he did not even coin the term "segregation" that is so often ascribed to him; or the phrase "independent assortment."

He did not discover dominant and recessive "alleles" or "versions of the gene" or "factors" or anything else. In fact, unlike most present day elementary textbooks, he actually used "dominant" and "recessive" quite correctly to describe the *appearance* of the character—what we would nowadays call the phenotype. Contrary to what is said in almost any elementary textbook, dominance and recessiveness are not features of the genes; they are features of the *phenotype*, i.e., the *expression* of the genes. That is exactly how Mendel used the terms.

At a more mundane level, he never used letters derived from the external appearance of the character to represent anything. For example, the "tallness" was never shown as T and the "dwarf" was never t. "Smooth pea" was not S and the "wrinkled pea" was not s. He *did* use upper case for the dominant character (i.e., the dominant *phenotype*) and lower case for the recessive, but invariably the letters Mendel used were A and a. If he was considering a second character at the same time as the first, he would use the next letter in the alphabet to represent the second. That is, B and b. And C and c for a third, and so on. Mendel tended to use these letter symbols to represent the phenotypic appearance of the plant, not the "things," factors, or "genes" that we now know are the underlying *causes* of that appearance. Thus in his world, A is a pure breeding tall plant; and a is a pure breeding dwarf plant, and a hybrid between the two, the result of artificially crossing one with the other is Aa. Throughout this chapter I have endeavored to use the same notation as Mendel did in order to avoid putting a twentieth-century gloss on the account.

In a way, this question of notation is the nub of the problem: Mendel had no idea about genes, so he could not really give them symbols. At his time biologists had only just come to terms with the existence of *cells* and they were still arguing about the meaning of the *nucleus*. It is no wonder that Father Gregor didn't give symbols to something that had not been discovered—but sometimes, just occasionally in the course of his beautiful, lucid, exemplary explanations, he lets the rigorous, objective reasoning slip. For an instant he appears to be talking about something that we would today call "genes." For example, at one point when he is talking about the crossing of two hybrid plants, **Aa** x **Aa**, he writes:

> It remains, therefore, purely a matter of chance which of the two sorts of pollen will become united with each separate egg cell. According, however, to the law of probability, it will always happen, on the average of many cases, that each pollen form **A** and **a** will unite equally often with each egg cell form **A** and **a**, consequently one of the two pollen cells **A** in the fertilization will meet with the egg cell **A** and the other with the egg cell **a**, and so likewise one pollen cell **a** will unite with an egg cell **A**, and the other with the egg cell **a**.

Pollen cells: A   A   a   a

Egg cells: A   A   a   a

> The result of the fertilization may be made clear by putting the signs for the conjoined egg and pollen cells in the form of fractions, those for the pollen cells above and those for the egg cells below the line. We then have:

$$\frac{A}{A} + \frac{A}{a} + \frac{a}{A} + \frac{a}{a}$$

> In the first and fourth term the egg and pollen cells are of like kind, consequently the product of their union must be constant, namely **A** and **a**; in the second and third, on the other hand, there again results a union of the two differentiating characters of the stocks, consequently the forms resulting from these fertilizations are identical with those of the hybrid from which they sprang. *There occurs accordingly a repeated hybridization.* [Mendel's emphasis] This explains the striking fact that the hybrids are able to produce, besides the two parental forms, offspring that are like themselves. **A/a** and **a/A** both give the same union **Aa**, since, as already remarked above, it makes no difference in the result of fertiliza-

tion to which of the two characters the pollen or egg cells belong. We may write then $A/A + A/a + a/A + a/a = A + 2Aa + a$.[6]

In this passage one can see how Mendel's notation came under strain. A is the pure breeding dominant pea plant, and a is the pure breeding recessive, but *just for a moment* A and a have become symbols for something very different: for a few lines they represent what is *inside* the pollen cells and the egg cells, that is, the genes (or alleles). He even shows them together making up the offspring, as a fraction. A/A or A/a or a/a. Then he goes back to his original notation, where the pure breeders are A or a and only the hybrid forms have two letters, Aa.

This is the key to all debate about what Gregor Mendel thought he had discovered. Elsewhere he betrays his own difficulty with the problem, making it quite clear that he understood the fundamental difference between the outward appearance of a plant and the "internal composition" of the cells.[7] He confronts the issue in the concluding remarks to the paper:

> In the opinion of renowned physiologists, for the purpose of propagation one pollen cell and one egg cell unite in seed-bearing plants[8] into a single cell, which is capable by assimilation and formation of new cells to become an independent organism. This development follows a constant law, which is founded on the material composition and arrangement of the elements which meet in the cell in a viable union. If the reproductive cells be of the same kind and agree with the foundation cell of the mother plant, then the development of the new individual will follow the same law which rules the mother plant. If it chances that an egg cell unites with a *dissimilar* pollen cell, we must then assume that between those elements[9] of both cells, which determine opposite characters some sort of compromise is effected. The resulting compound cell becomes the foundation of the hybrid organism the development of which necessarily follows a different scheme from that obtaining in each of the two original species.

If, in this passage, you substitute the word "gene" for "element" you have it. What he had seen, dimly because he was staring into the unknown, was that inheritance of characteristics was under the control of things, particles, elements that were passed from the parent into the sex cells and so to the offspring; and that there were two of these "elements" for each basic character, one contributed by the father and the other by the mother. With this passage Gregor Mendel was shining a light into the darkness ahead: he had actually understood how inheritance worked. His tragedy was that there was no one to step forward with him.

XVII, 3.   *106. Leguminosae.*

454. Phaseolus coccineus L.   **Feuer-Bohne.**

The Follow-up

So the experimental work came to an end—ironically his plants were finally destroyed by his old enemy the pea weevil—and in 1865 Mendel delivered his two lectures to the Brünn Society for Natural Science. What did the members make of it? There must have been applause. Anyone could see that what he had done was remarkable; sadly, they just couldn't see what it all meant. One of the difficulties was that they thought the whole lecture had been about hybridization between different species, and that is hardly surprising because Mendel himself thought it was about hybridization between closely related species. In fact that is what he *called* it: *Versuche über Pflanzen-Hybriden* (*Experiments in Plant Hybridization*). So it is hardly surprising that everyone in the audience made that mistake. They, like Mendel himself, were only children of their time, and at the time hybridization between species was thought to be a possible mechanism for evolution. Franz Unger thought as much, and he had been one of Mendel's lecturers at Vienna University. But, as Mendel shows so clearly in the introduction to the paper, even he couldn't really distinguish between species and variety in the garden pea, and anyway, he said, it didn't matter. It doesn't matter, because s*pecies* hybridization isn't actually what the work was all about: whatever he thought he had done, in fact Father Gregor had investigated hybridization *between close members of the same species*, varieties that differ only in one or two very specific, clear-cut characteristics.

At the end of the final lecture there were some comments and some questions, and later satisfactory mentions in the local newspapers—you may imagine the kind of thing: "Father Gregor Mendel, whom many readers will know as a popular and effective teacher at the *Realschule*, last night delivered a fascinating lecture on hybridization between different varieties of garden pea . . ."

And then nothing, until a few months later the secretary of the society, von Niessl, asked him to write up the report for publication in the *Proceedings* of the Society. So Mendel went back to his lecture notes, checked them against the extensive records that he had made of his experiments, and then began, in his meticulous copperplate, to write up the fair copy.

Scrawled diagonally across the top left hand corner of the

To confirm his findings in the peas, Mendel conducted hybridizing experiments in other species, such as the scarlet runner bean (*Phaseolus coccineus*).

RIGHT: When Mendel wrote this manuscript in 1865, the results of eight years of experiments received little notice. This paper is now considered a key publication in the history of science.

OPPOSITE, TOP: An offprint of *Experiments in Plant Hybridization* by Gregor Mendel. At his own expense, Mendel distributed forty copies of his paper, which was published in the proceedings of the Natural Science Society of Brünn in 1866. This journal's low profile may have hindered the impact of Mendel's results on the scientific community.

OPPOSITE, BOTTOM: Carl von Nägeli, a German botanist who was well known for his theories of inheritance, would seem an ideal candidate to review Mendel's results, yet he completely failed to understand Mendel's findings.

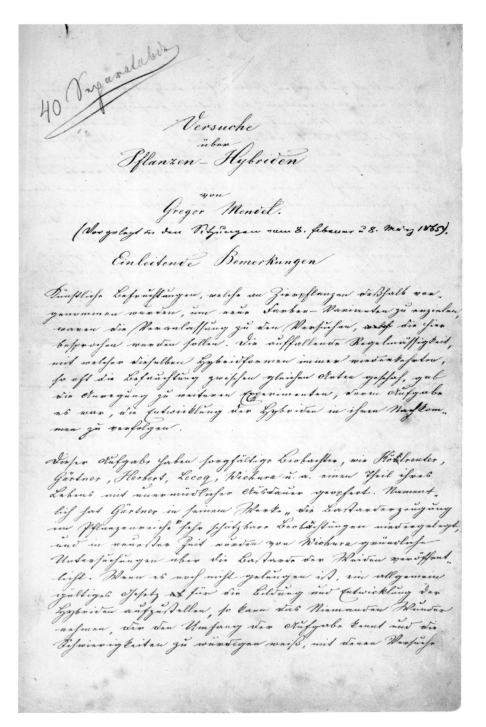

Gregor Mendel: Planting the Seeds of Genetics

manuscript, in the hand of von Niessl, is the phrase *40 Separatabdrucke*—"forty offprints." There is a mystery, both about the original document and those forty copies.

The Manuscript

When Iltis wrote his biography of Mendel in the early 1920s, he had the manuscript version of Mendel's paper on his desk in front of him. He himself had rescued it from a box of documents about to be destroyed; later he gave it for safekeeping to the Brünn Society for Natural Sciences, which was still in existence at the time. For years it was held in a bank vault in the city and taken out so that a photographic facsimile could be made in 1940. From there it disappeared. Long thought to have been destroyed, in the 1980s it turned up in the possession of a descendant of one of Gregor Mendel's sisters. Although various people have had access to the document, for the moment this historical treasure remains hidden from public view.

The Offprints

Whether all forty offprints were dispatched by Mendel to suitable people, we do not know. Only seven[1] of them have ever come to light:

On December 31, 1866, Mendel himself sent one copy to Carl von Nägeli, Professor of Botany at Munich University with a cover letter. This is the one copy that had any significant consequence. The very next day, on January 1, 1887, Mendel dispatched a further a copy to Kerner von Marilaun, who had been one of his lecturers at Vienna University but was now Professor of Botany at Innsbruck. When it was found again after the "rediscovery" of Mendel's work, this reprint was still uncut—that is, the pages, fan-folded when the booklet was bound, had never been cut open—which means that no one ever read it.

The third known offprint is in the Institute of Botany of the University of Amsterdam. It is this copy that landed on Hugo de Vries's desk in 1900 and prompted him to cite Mendel's work in his own paper—the paper that truly opened the world to modern genetics. The story of this will come later (see pp. 96ff.).

The fourth copy was found in the library of the Institute of Botany at Graz University. It is likely that this copy was actually sent by Mendel to his one-time

Versuche

über

Pflanzen-Hybriden,

von

Gregor Mendel.

(Separatabdruck aus dem IV. Bande der Verhandlungen der naturforschenden Vereines.)

Im Verlage des Vereines.

Brünn, 1866.

Aus Georg Gastl's Buchdruckerei, Postgasse Nr. 446.

Vienna lecturer, Franz Unger, who had by then retired. This copy was also uncut.

A fifth copy is in the library of Indiana University, having got there via Charles Davenport, who obtained it about 1898.

The sixth offprint was discovered in the archive of the Brünn Natural History Society and a seventh copy was found in the abbey library itself. Of the remaining thirty-three offprints not a trace remains. Possibly one or two copies have yet to come to light but one must accept that the majority are lost forever. Certainly, for all the reaction they achieved, they might as well never have been sent in the first place.

The Proceedings

The 1866 Proceedings of the Brünn Society of Natural History, including Mendel's paper, were duly published. One hundred twenty copies were distributed to various learned societies around the world, including the Linnean Society and the Royal Society in London, universities and libraries in Berlin and Vienna, and four universities in the U.S. No one took much notice, although the paper by Mendel was cited in a number of books, particularly W. O. Focke's monumental work *Plant Hybridization*[2] of 1881. In this book there are some fourteen references to Mendel's work, some to the paper on the garden pea, most to the later paper on *Hieracium*. The most enlightening of the passages in Focke will give a flavor of how a nineteenth-century biologist would have seen Mendel's remarkable work, and how he would have failed to grasp the importance:

> Mendel's numerous crossings gave results which were quite similar to those of Knight, but Mendel believed that he found constant numerical relationships between the types of the crosses. In general, the seeds produced through a hybrid pollination preserve also exactly the color which belongs to the mother plant, even when from these seeds themselves plants proceed, which entirely resemble the father plant, and which then also bring forth the seeds of the latter.[3]

This was unlikely to stir the reader to rush to the original work, but even had he done so, would he have recognized its importance? That is a question that is still debated. Ironically, Focke's bibliography, including the Mendel reference, was quoted by a number of authorities, including, indirectly, Charles Darwin. He lent his copy of Focke's book to his disciple George Romanes who was preparing an article on hybridization for the Encyclopaedia Britannica (ninth edition). Romanes duly lifted the bibliography from Focke's book, and therefore cited the paper *Versuche über Pflanzen-Hybriden* by Gregor Mendel. However, it is certain that neither Romanes nor Darwin actually *read* the part in Focke's book that refers to Mendel's paper because the book is still in the Darwin library, with the relevant pages still uncut. It is probably this fact that has led to the entirely fictional story that still occasionally appears in the literature, that Darwin was actually sent Mendel's paper but never read it.

In fact, almost a dozen scientific works before 1900 refer to Mendel's paper but none of them bore any fruit and only one follow-up to his work was of any significance. That was the personal contact that he made with Carl von Nägeli.

The Letters to Carl von Nägeli

Carl von Nägeli was Professor of Botany at Munich University and a man of great reputation as one of the principal theorists about the mechanisms of inheritance. It was to this august personage that Mendel addressed his first offprint. The letter is dated January 31, 1866. "Highly esteemed sir!" it begins. "The acknowledged pre-eminence your honor enjoys in the detection and classification of wild-growing plant hybrids makes it my agreeable duty to submit for your kind consideration the description of some experiments in artificial fertilization."

The correspondence with von Nägeli was little short of disaster, and yet, ironically, these ten letters are the only insight we have into Mendel's ideas

These notes from around 1870 are among the only existing records of Mendel's experiments. Here, Mendel compares the results obtained with peas and hawkweed to those obtained with willow hybrids.

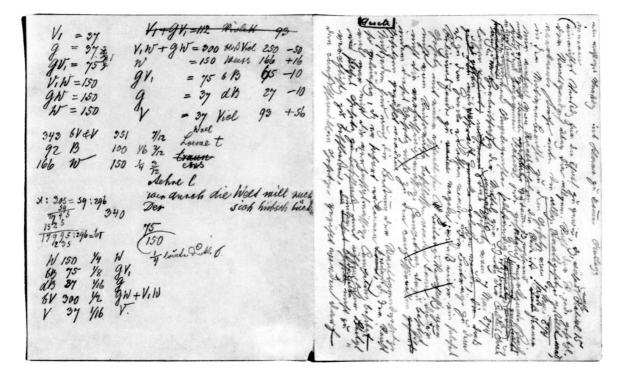

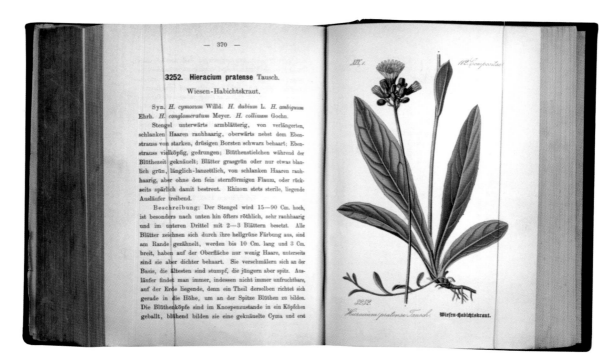

3252. Hieracium pratense Tausch.

Wiesen-Habichtskraut.

Syn. *H. cymosum* Willd. *H. dubium* L. *H. ambiguum* Ehrh. *H. conglomeratum* Meyer. *H. collinum* Gochn.

Stengel unterwärts armblätterig, von verlängerten, schlanken Haaren rauhhaarig, oberwärts nebst dem Ebenstrauss von starken, drüsigen Borsten schwarz behaart; Ebenstrauss vielköpfig, gedrungen; Blüthenstielchen während der Blüthezeit geknäuelt; Blätter grasgrün oder nur etwas bläulich grün, länglich-lanzettlich, von schlanken Haaren rauhhaarig, aber ohne den fein sternförmigen Flaum, oder rückseits spärlich damit bestreut. Rhizom stets sterile, liegende Ausläufer treibend.

Beschreibung: Der Stengel wird 15—90 Cm. hoch, ist besonders nach unten hin öfters röthlich, sehr rauhhaarig und im unteren Drittel mit 2—3 Blättern besetzt. Alle Blätter zeichnen sich durch ihre hellgrüne Färbung aus, sind am Rande gezähnelt, werden bis 10 Cm. lang und 3 Cm. breit, haben auf der Oberfläche nur wenig Haare, unterseits sind sie aber dichter behaart. Sie verschmälern sich an der Basis, die ältesten sind stumpf, die jüngern aber spitz. Ausläufer findet man immer, indessen nicht immer unfruchtbare, auf der Erde liegende, denn ein Theil derselben richtet sich gerade in die Höhe, um an der Spitze Blüthen zu bilden. Die Blüthenköpfe sind im Knospenzustande in ein Köpfchen geballt, blühend bilden sie eine geknäuelte Cyma und erst

Hieracium pratense Tausch. **Wiesen-Habichtskraut.**

Mendel could not duplicate his pea experiment results on hawkweed (*Hieracium*). Almost forty years later it was discovered that most species of hawkweed can produce seeds without fertilization, so many or most of the plants Mendel thought he was carefully fertilizing were reproducing themselves on their own.

following the publication of his paper. From the start there is a condescending tone in the professor's responses to the insignificant little friar from Moravia, but what is far, far worse than the manner of his replies is the lack of comprehension that he displays. This was the man who should have been amazed by Mendel's achievement. Instead we get, in von Nägeli's first reply, the deflating comment that "it seems to me that the experiments with *Pisum*, far from being finished, are only beginning . . ."

There is worse to come. In his first letter Mendel had mentioned the possibility of repeating his hybridization experiments using the hawkweeds, genus *Hieracium*, plants that were von Nägeli's special area of interest. For all the professor's grandeur, the hawkweeds are a common lot. To most people the name "weed" seems appropriate. Their flowers resemble dandelions and like dandelions they are members of the Compositae, the daisy family, which means that the "flowers" aren't actually single flowers at all but "composites" of dozens of minute florets. The manipulation of such tiny floral parts in the way that Mendel had manipulated pea flowers is almost impossible. Nevertheless, following von Nägeli's encouragement, Mendel tried. He almost ruined his eyesight doing it and we now know that the effort was pointless anyway: the *Hieracia* are apomictic. This means that they set seed without fertilization and therefore offspring are, in effect, clones of their parent. Cross-breeding with such plants is impossible. Mendel was not to know this (it wasn't discovered until the next century),

Gregor Mendel: Planting the Seeds of Genetics

but it effectively made his attempt to cross-breed different hawkweeds totally meaningless.

In all fairness von Nägeli cannot be blamed for not knowing that *Hieracium* is apomictic, and maybe he cannot be blamed for his failure to see that this strange correspondent from Moravia had just unlocked the door to a whole new science. He was, like anyone else, a child of his times, and the general current of scientific thought was against Mendel. But von Nägeli can surely be blamed for lines like this:

> I refrain from comment upon other points in your communication, for, without detailed knowledge of the experiments upon which your monograph is based, what I can say would be mere surmise.[4]

Without detailed knowledge? He had just been given a concise summary of the most complete piece of quantitative experimental work that had ever been performed in biology, and he refrains from making a comment because he doesn't have detailed knowledge. Or this:

> You should regard the numerical expressions as being only empirical, because they can not be proved rational.[5]

How can anyone have even glanced through Mendel's paper and concluded that there is no rational basis for the numerical expressions that he has derived from his results? The very thing that they *are* is rational. Rationality is what defines them. In response to criticisms of this kind, von Nägeli gets the full Mendel treatment in the latter's second letter (April 18, 1867)—an insistence that he *has* moved from the purely empirical to the rational, a carefully reasoned argument in support of this, details of different types of crosses he has made, and meticulously labeled packets of dried peas for von Nägeli to plant so that he too may see with his own eyes (in fact, von Nägeli had *asked* Mendel for these, but of course, he never planted them).

So Mendel gathered the hawkweeds and tried artificial crosses and wrote with enthusiasm of what he was planning; but we know that he was attempting the impossible, straining his eyes to see floral parts that are almost microscopic, trying to cross-pollinate plants that don't bother with pollination at all. In fact he was setting off down a dead-end path, lured back to the old nineteenth-century ways of hybridizing plants from different species and without clearly defined characteristics to "see what happens."

The second and third letters in the series date from spring and winter of

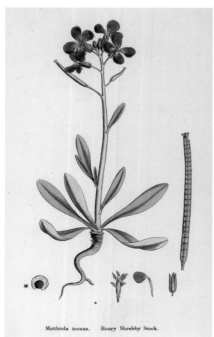

Stock (*Matthiola incana*)

1867 and are concerned with botanical matters, but by May 1868 Mendel is giving some different news from the abbey:

Recently there has been a completely unexpected turn in my affairs. On March 30 my unimportant self was elected life-long head, by the chapter of the monastery to which I belong. From the very modest position of teacher of experimental physics I thus find myself moved into a sphere in which much appears strange to me, and it will take some time and effort before I feel at home in it. This shall not prevent me from continuing the hybridization experiments of which I have become so fond; I even hope to be able to devote more time and attention to them, once I have become familiar with my new position.

"Your devoted friend," he signs off now, "Abbot and Prelate of the Convent of St. Thomas."

Of course he was wrong about having more time for his experiments. His appointment to one of the most powerful positions in the Moravian Church *did* affect his ability to pursue his experimental work. In the summer of 1868 he explains to von Nägeli that he has to leave Brünn on an inspection of the convent's properties. He leaves strict instructions with the gardener how to pot hawkweeds that von Nägeli has sent, but on his return finds that they have been over-watered and all of them are dead. So, one feels, is the magisterial piece of work that he has done, along with the chance of his ever having it recognized.

Yet there are still moments in the correspondence when Mendel's genius shines through the difficulties and disappointment. As soon as he gets away from the accursed hawkweeds and back to the species of his choice there is all the clarity and perception of his earlier work. In the eighth letter to von Nägeli, he mentions the numerous experiments that he has done to confirm that his findings in the garden pea are reflected in other species, and he nods toward Darwin as he does so:

Of the experiments of previous years, those dealing with *Matthiola*[6] *annua*. and *glabra*, *Zea*, and *Mirabilis* were concluded last year. *Their hybrids behave exactly like those of Pisum.*[7] Darwin's statements concerning hybrids of the genera mentioned in "The Variation of Animals and Plants under Domestication," based on reports of others, need to be corrected in many respects. Two experiments are still being continued. I have about 200 uniform specimens of the hybrid of *Lychnis diurna* and *L. vespertina.* The first generation should flower in August.

And still von Nägeli cannot see the importance of his correspondent's work! Mendel has found a rule that governs the inheritance of clearly defined characters in one species, and now he has confirmed that it holds true for other species in other genera, and still he is ignored. Yet Mendel goes on. For all his distractions as abbot he is still worrying away at the problems that he has set himself. In the following extract it is worth noticing his remarkably prescient suggestion about the probability of competition between pollen grains during fertilization—with its strong echoes of Darwinian natural selection:

Four O' Clock
(*Mirabilis jalapa*)

Because of my eye ailment I was not able to start any other hybridization experiments last year. But one experiment seemed to me to be so important that I could not bring myself to postpone it to some later date. It concerns the opinion of Naudin and Darwin that a single pollen grain does not suffice for fertilization of the ovule. I used *Mirabilis jalappa* for an experimental plant, as Naudin had done; the result of my experiment is, however, completely different. From fertilizations with single pollen grains, I obtained 18 well-developed seeds, and from these an equal number of plants, of which ten are already in bloom. The majority of the plants are just as vigorous as those derived from free self-fertilization. A few specimens are somewhat stunted thus far, but after the success of all the others, the cause must lie in the fact that not all pollen grains are equally capable of fertilization, and that furthermore, in the experiment mentioned, the competition of other pollen grains was excluded. When several are competing, we can probably assume that only the strongest ones succeed in effecting fertilization. It should also be possible to prove directly by experiment whether or not two or more pollen grains can participate simultaneously in the fertilization of the ovule in *Mirabilis.* According to Naudin, at least three are needed![8]

It is important to grasp what he has done here: working with the microscope, he has transferred *single pollen grains* from the anther to the stigma and then grown viable seed from the resulting fertilization. Although this is remarkable it is not quite unique (Kölreuter had performed a similar experiment on *Mirabilis*), but Mendel's proposed *experiment* certainly is unique: in the next letter[9] he gives details, explaining that he has carried out simultaneous pollinations with *two* grains of pollen, one from a pure-bred yellow flower, the other from a pure-bred white flower, both placed on the stigma of a pure-bred red flower. Already knowing the result of *red flower x yellow flower* and *red flower x white flower*, he is thus offering the plant a choice. Will it make a *red x yellow* cross or a *red x white* cross, or will *both* pollen grains somehow become involved and produce another, unexpected color, thus showing that two pollen grains can contribute to one fertilization? Like so much of his work, like so many of the greatest experiments, it is brilliantly simple and logical.

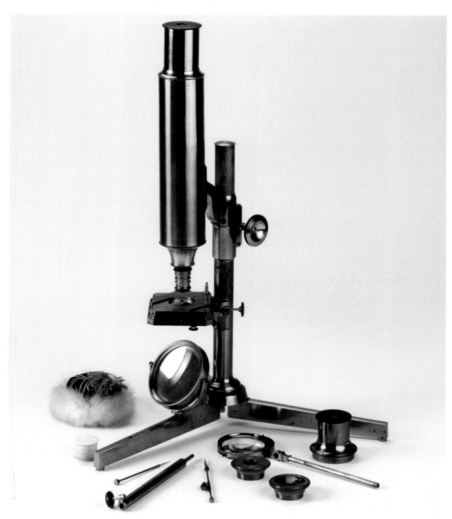

Mendel used his microscope to perform single-grain pollination, contradicting claims by Darwin and others that one grain of pollen was not enough to fertilize a plant. Mendel could see pollen magnified 179 times.

Gregor Mendel: Planting the Seeds of Genetics

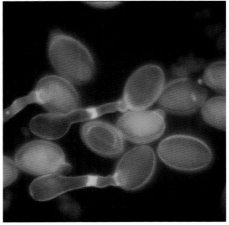

Sadly, we never hear the result of this experiment because there is a gap in the correspondence now. Mendel has too much to do, too many demands on his time, and there are always his meteorological observations and the astronomical observations (see p. 85), and perhaps he grew tired of the incessant demands of hybrid plants and those miserable hawkweeds that would not follow the rules. Only after a pause of three years does he contact von Nägeli once more, and it is by way of being a valediction:

> Despite my best intentions I was unable to keep my promise given last spring. The *Hieracia* have withered again without my having been able to give them more than a few hurried visits . . .[10]

And that is it. With a passing reference to his difficulties with the hawkweeds and the expression of his greatest admiration and esteem he signs himself, very respectfully, Gr. Mendel, and the correspondence is at an end. Never has a greater scientific opportunity been so comprehensively missed.

The Anlage

However, the Mendel-von Nägeli correspondence cannot be left at that, because in the penultimate letter there is a passage of great significance in understanding Mendel's own ideas about inheritance. Popular textbooks credit him with having "discovered the gene"—or at least grasped the fact that inheritance is in some way "particulate"—but ever since the rediscovery of his work in 1900 there has been debate about this in learned journals. Some authors have called into doubt whether he really *did* grasp the idea of a "particle of inheritance" in any meaningful way, and rather than call him the first geneticist they put him into the strictly nineteenth-century world of plant hybridists. Certainly that is where he came from, but is it where he ended up—or was he truly ahead of his time?

In the letter to von Nägeli of September 1870, Mendel makes a speculation about the inheritance of sex in *Lychnis* species. This passage is often overlooked, possibly because the speculation itself is erroneous, but it carries with it the answer to the question that so many biologists and historians of science have asked—did Mendel really understand what he had discovered? And further— did he really predict the existence of things that we now call genes? I quote it for that:

Finally, let me report on a curiosity in the numerical ratios in which the male and the female plants of the hybrid *Lychnis diurna+L. vespertina occur.* I fertilized three flowers of *L. diurna* and planted the seeds of each capsule separately. They produced:

capsule 1:	74	plants of which	54	female	20	male
capsule 2:	58	plants of which	43	female	15	male
capsule 3:	71	plants of which	54	female	7	male
	203		151		52	

Is it chance only that the male plants occur here in the ratio 52/203 or ¼, or has this ratio the same significance as in the first generation of hybrids with varying progeny? I should doubt the latter, because of the strange conclusions which would have to be drawn in this case. On the other hand the problem can not be so easily dismissed if one considers that the *Anlage* ("factor") for the functional development of either the pistil (female part) alone or of the anthers (male parts) alone, must have been expressed in the organization of the primordial cells from which the plants developed, and that this difference in the primordial cells could possibly be due to the ovules as well as the pollen cells being different as regards the sex *Anlage*. Therefore I do not want to dismiss the matter completely.

This is yet another extraordinary piece of work. *Lychnis* are dioecious plants, that is, they have separate sexes, but Mendel is wrong with his explanation of how their sex is determined. He suggests that this is another example of a hybrid cross, **Aa** x **Aa**, with the males being the ¼ of the offspring that are recessive. He himself sees the problems attached to such an explanation, and we now know that in the campions sex is determined by an XY chromosome system exactly as in humans. However, he is perfectly correct in discovering a strong female bias in the sex ratio. In 1928 Correns (stimulated by a reading of this letter to von Nägeli?) investigated this and it has been the subject of research ever since. No clear explanation has yet been found, although one of the suggested explanations has been the superior competition by X-carrying pollen grains. Mendel would certainly have understood the idea of competition among pollen grains—see the letter of July 3 quoted above—but what he would *also* have understood is the later concept of a "particle of inheritance," a "gene"—because there it is in this passage. In all Mendel's writing this is only the second occasion in which he uses the term *Anlage*. In the paper on the garden pea, except for a single passing use toward the end, he invariably talks about *Merkmal* (character) or uses the borrowed English word "character" or, in a significant passage referred to in the preceding chapter, the word *Element*. But in the ninth letter to von Nägeli he is using this new word—*Anlage* (factor)—to describe that thing that causes, in this case, the development of sex of the plant in question. The *Anlage* resides, he says, in the primordial cells from which the plant develops, and we know and he knows that it can only have got there by being carried in the pollen and the egg cell. It is what we would nowadays call a gene.

There is a footnote to the sorry story of Mendel's correspondence with von Nägeli. At the time he was writing to the friar from Moravia, von Nägeli must have been preparing his great work entitled *A Mechanico-Physiological Theory of Organic Evolution* (published in 1884, the year of Mendel's death) in which he proposes the concept of the "idioplasm" as the hypothetical transmitter of inherited characters. The book dwells at length on crosses between different varieties of plants and animals, and describes the effect of dominance and the reappearance of recessive parental types when hybrids are inbred. Yet in it there is *not a single mention* of the work of Gregor Mendel. We can forgive von Nägeli for being obtuse and supercilious. We can forgive him for being ignorant, a scientist of his time who did not really have the equipment to understand the significance of what Mendel had done despite the fact that he (von Nägeli) speculated extensively about inheritance. But omitting an account of Mendel's work from his book is, perhaps, unforgivable.

Paraguay (Port Maldonado)

Brésil

Life beyond Peas

In the popular image of the reclusive friar pottering around his monastery garden, the fact of war seems quite out of place. And yet in 1866, just as Mendel's paper was being prepared for publication, the Prussian army invaded the Austro-Hungarian Empire. The reasons for this brief (it is known as the Seven Week War) but significant irruption lie in the whole question of developing German identity at the time, and the rivalry between Prussia and Austria for influence over the political mess of Germany. This rivalry would continue throughout the nineteenth and twentieth centuries and culminate in the absorption of Austria into the Nazi's "greater Germany" in 1938.

On this early occasion Prussian troops, commanded by the formidable Helmuth von Moltke, invaded the Empire from the north in a three-pronged attack that converged on the defending army at Königgrätz (modern Hradec Králové). Here, on July 6, the Prussians inflicted a humiliating and crushing defeat on the Austrians. Moving south toward Vienna five thousand Prussian troops then occupied the city of Brünn on July 12. Vienna, the very heart of the Empire, was less than a hundred miles away.

One can imagine the citizens of Brünn watching with dismay as the invaders paraded along the wide esplanade of the Ringstrasse, pennants and banners flying, pipes and drums playing. Within living memory the French army had occupied the city and Napoleon himself had spent time in the Spielberg, the castle that glowers over the town, before moving on to win the battle of Austerlitz just to the east of the city. And now they were once again under a foreign army, but this time with French voices replaced by the accents of northern Germany. The king of Prussia himself moved into town, along with his chief minister, Otto von Bismarck.

Mendel saw all this from the convent below the Spielberg. He watched cavalry being led into the convent gardens, ninety-four horse with riders and grooms who had been billeted on the Augustinians. He watched sixteen officers take up quarters in another part of the building. The convent had to deal with them, the kitchens had to feed them, and all costs had to be met by the order. When those troops moved out

Mendel's attempts at bee-breeding included South American bees like these species.

After becoming abbot of the monastery in 1868, Mendel had less time for science. He may have been disheartened by the lack of reaction to his pea paper, but he knew that his discovery was important, and not long before his death in 1884 he told a colleague, "My time will come."

others took their place. Over the days fear and panic in the city were transformed into resentment and anger. Reports came in from the countryside of the army living off the land, expropriating animals and crops, tramping into cottages and demanding accommodation, and driving the owners out into the stables. And with the soldiers came cholera. For six weeks the disease seethed in the overcrowded city. The hospital just up the road from the convent filled with victims. A thousand civilians died, along with two thousand of the invaders. Some of the friars got mild attacks; Father Anselm's mother, a resident of the convent, died.[1]

Relief came after a few months when a truce was signed at nearby Nikolsburg (Mikulov) and the Prussians finally withdrew, leaving the citizens to reconstruct their lives. Mendel tended what was left of his plants, took his weather

Gregor Mendel: Planting the Seeds of Genetics

readings, measured the water level in the convent well (it was unusually low that year), wondered about the connection between subsoil water level and disease epidemics, and about sunspots, weather patterns, and the curious breeding habits of bees. In the autumn the Proceedings of the Natural Science Society were published and the forty offprints of his paper delivered to him. And that winter he sat down to write to Carl von Nägeli. Life—teaching, checking his results, trying new crossings with other plant species, the daily round of the convent, the meteorology, the bees—went on.

Abbot

In March 1868 Abbot Napp, who had been ill for some time, died. The community convened in the chapter house of the convent to elect his successor and after two inconclusive rounds of voting, Gregor Mendel was elected. The final vote wasn't quite unanimous—to the very end Mendel himself voted for his radical friend, Matouš Klácel.

As he wrote to von Nägeli, "Thus, all at once, from the very modest position of teacher of experimental physics I thus find myself moved into a sphere in which much appears strange to me, and it will take some time and effort before I feel at home in it."

He took his duties as abbot seriously, but his other interests were never totally lost. Throughout his adult life he was a meticulous collector of scientific data, not just plant and animal but also astronomical and meteorological. Three times every day (at 6 A.M., 2 P.M., and 10 P.M.), come rain, come snow, he would make the round of his instruments situated in various parts of the monastery grounds. His weather records are still extant, pages and pages of temperatures and pressures, of wind speeds and hours of sunlight, even depth of the water table, all entered into leather-bound ledgers in his immaculate, copperplate hand. The records are vivid witness to his method and accuracy, those qualities that were so important in his breeding experiments. For many years he reported to the Central Meteorological Institute in Vienna, and he was one of the founding members of the Austrian Meteorological Society. He continued with this ritual of recording the data until the last month of his life.

Then there were the astronomical observations. Mendel's telescope, adapted for sunspot observation, is still in existence, as are some of his sunspot drawings. It seems likely (Iltis quotes von Niessl as confirming this) that Mendel was attempting to discover some correlation between weather

The abbey telescope. Mendel was well versed in probability theory—the mathematical study of probability—and combinatorial mathematics (used to calculate the ways that certain patterns can be formed). These theories grew directly from astronomy in the 1800s, and Mendel applied them to his plant studies.

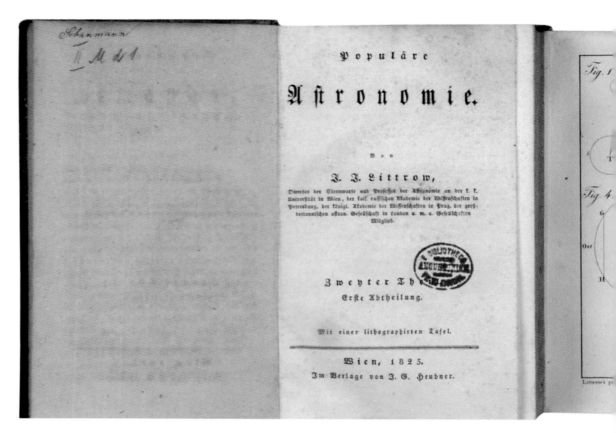

Popular Astronomy by Joseph Johann Littrow. Besides being a general guide, this book promotes the role of probability theory in astronomy. This is Mendel's own well-marked copy.

phenomena and sunspot activity—always his mind was working, hypothesizing, attempting to uncover patterns in the workings of nature.

Tucked away on the steep hillside immediately behind the convent, the red brick bee-house that Mendel had built can still be seen. It was here that he pursued another of his scientific interests, the artificial breeding of bees using a cage that he had specially built by a local carpenter. Although the project never came to much—bees are notoriously strange in their breeding habits—it is interesting to note such clear evidence that Mendel was also interested in the genetics of animals. We know from a passing mention in Iltis that he kept mice in his rooms in the convent and there seems to be little doubt that he was interested in their patterns of inheritance. Iltis also reports that Mendel's nephew, Ferdinand Schindler (a doctor), told him of his uncle's interest in human inheritance, even making a pedigree of his own family showing such characters as height, hair color, and baldness (Mendel himself was balding). Schindler was one of the three sons of Mendel's sister, Theresia, who had helped him out with money when he was a struggling student at Olmütz. Mendel never forgot his sister's kindness, and was not only a friend and confidant to her sons throughout his life, but also supported them financially in their studies.

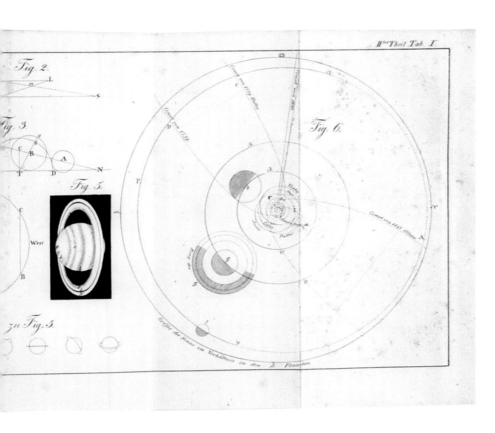

Mendel's bee house at the abbey. Around 1871 Mendel began cross-breeding bees, possibly attempting to confirm the results of his pea experiments.

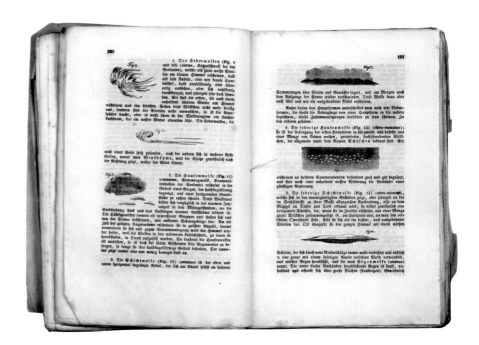

Mendel's copy of *Textbook of meteorology presented in easily understood form* by August Kunzek. Mendel's plant-breeding experiments were rooted in the same principles as his meteorology studies: the observation of natural phenomena and the use of statistics to understand changes.

The last years of his life are marked by a dogged, obsessive campaign to protect the convent from the tax demands of the government. It is the counting and recording of garden peas translated into local politics. Toward the end, this battle became so intransigent that the other members of the order considered having him declared unfit to govern them. But before that sad decline he was a pillar of the community, and, in a small but significant way, a sponsor of the arts. In 1872 he received a certificate of honor from the Brünn Musical Society, but his greatest contribution to music was quite different and probably not appreciated by Mendel himself: in that same year he appointed as organist and choirmaster a young man called Leoš Janáček, who had previously been at the choir school of the abbey. In later years Janáček recalled the convent with distinct unhappiness: "the gloomy corridors, the old church, the gardens, my poverty-stricken youth, my loneliness, my homesickness." Obviously he didn't appreciate that in those gardens the first steps in a new science were being taken. But he did go on to become the third great Czech composer, alongside Smetana and Dvořák. His last opera, *The Cunning Little Vixen*, tells the story of a female fox reared in captivity by a forester; and indeed, at the time that Janáček was a small boy in the choir school, Father Gregor Mendel himself kept a pet vixen in the convent garden.

The Tax Man Cometh

There is evidence—a scrap of notes found among official correspondence and probably dating from somewhere between 1875 and 1880—that Mendel continued to mull over his ratios to within a few years of his death, but substantially his genetics work was over by the end of his letters to von Nägeli. Bereft of colleagues who might have grasped his ideas and thus provided a sounding board for further speculation, frustrated by the hawk-weeds[2] and the disappointing response from von Nägeli, he had lost his optimism about his discoveries. "My time will come," he is famously reported to have said to von Niessl, but perhaps the statement has more the flavor of defiance than conviction.

These notes from around 1870 are among the only existing records of Mendel's experiments. Here Mendel recalculates data from later experiments in plant hybridization.

At the start of his abbacy he was a genial, popular man, given to sharing a beer with the locals (the Starobrno brewery that still stands beside the convent was owned by the abbey in those days) and playing skittles with such notables as the Lord Lieutenant of Moravia and the president of the supreme court. But what really occupied him in the last decade of his life was that bitter and, in true Mendel fashion, obsessive row with the government over new tax demands on the abbey's properties. Ultimately, it led to his isolation from public life and alienation from his fellow friars. In the end it was resolved as such rows often are—by the death of one of the combatants and accommodation on the part of his successor.

During these later years Mendel suffered from various degenerative illnesses. He was overweight, a heavy smoker, and subject to hypertension and progressive kidney failure. Toward the end his body was swollen by fluid retention—symptomatic of a failing heart. He kept up his ironical good humor to the end, in December 1883 writing to one of his pupils who had become a meteorologist: "Since we are not likely to meet again in this world, let me take the opportunity of wishing you farewell, and of invoking upon your head all the blessings of the meteorological deities." Even here, as throughout his extant writings, there is no expression of overt piety and no reference to the God of his religion.

He died during the early morning of January 6, 1884, and was buried three days later. A requiem mass was celebrated for him in the abbey church, the music being conducted by the former pupil at the abbey choir school whom he himself had appointed organist—the future composer Leoš Janáček. After his burial in the section of the public cemetery reserved for brothers of the Abbey of St. Thomas, his books were placed in the library of the convent and his official letters in the archive. His personal papers, his notes, and all the records of his breeding experiments were taken out to the hill behind the convent and burned. There was no indication that one of the greatest scientists of the nineteenth century—of any century—had died.

S. Holden, del. & Lith.

Antirrhinum majus flore pleno.

The Rediscovery

Why did the world not recognize Mendel's work as we now do? Partly it was simply that the scientific climate of the time was not ready to accommodate his new ideas or the implications behind them. Scientific discovery is often a matter of language. Sometimes—particularly in physics—it is the language of mathematics, but in biology it is usually the language that we employ in everyday communication.

Pangenes and Idioplasm

You cannot easily talk about a gene unless you have the word for it, and the word corresponds more or less exactly to something that is actually *there*. On the other hand, it is difficult to discover something that is actually there until you can talk about it. So scientists fumble around with half-expressed ideas and half-formed speculations and often it is only gradually that such ideas and speculations harden around a new concept, like calcium and carbonate ions hardening around an irritant grain of sand in an oyster to deliver up a pearl. So, during the second half of the nineteenth century, while Mendel abandoned his breeding and turned to the business of being abbot, and later, after his death in 1884, a whole collection of words gathered around the concept of inheritance. Von Nägeli wrote about the "idioplasm," Weissmann about the "germ plasm," Darwin about "gemmules," de Vries about "pangenes." Each word expressed an abstract idea, a more-or-less vague concept that lay within the inventor's mind. The problem lay in grounding the term in reality. Where exactly *was* the idioplasm, what *was* a gemmule? Where were you going to find such things? Could you put them on a microscope slide and observe them? The literal-minded searched and discovered nothing—but was that because the search was flawed, the instruments inadequate, or because the concept itself was wrong and the thing didn't actually exist?

Charles Darwin wrote about a hybridization experiment with the snapdragon (*Antirrhinum majus*) in his 1868 book, *The Variation of Animals and Plants under Domestication*. But in comparison to Mendel's thousands of pea plants, Darwin only grew two generations of snapdragon hybrids, with only 127 plants in the second generation.

Darwin's Pangenesis

Darwin's theory of pangenesis merits mention here not because it was correct—it was, in fact, nonsense—but because it was seriously considered at the time. The theory was invented

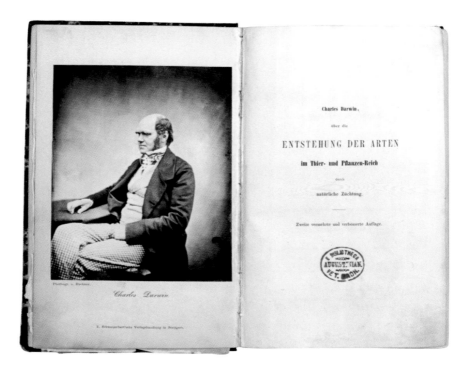

This is Mendel's own copy of *On the Origin of Species*. Darwin's theories were obviously an influence on Mendel's work, but there is no evidence that Darwin was aware of the friar's research.

because his book *On the Origin of Species* received a devastating review in an Edinburgh journal[1] by a Scottish engineer called Fleeming Jenkin. In this review Jenkin pointed out that the new Darwinian theory of natural selection foundered because there was no possibility of a new variation ever spreading through a population. The difficulty lay in the fact that the prevalent theory of inheritance available at the time was one of blending—that offspring were a "blend" of their parents' characteristics. Jenkin made the cogent point that given such blending, any rare novelty that arose by chance (in those days they were known as "sports"; nowadays we would call them mutations) would surely be swamped by all the normal types that made up the rest of the population. However useful it was, a new variation would simply be "blended out" as soon as the animal or plant bred with its fellows.

This logical point was fairly devastating and it hit its target. Darwin immediately addressed the problem and, in his next book, *Variation of Plants and Animals under Domestication* (1868), he came up with a theory of inheritance that was an alternative to blending: pangenesis. This theory was highly speculative and Darwin himself accepted that there was no experimental evidence for it. Nevertheless, he wrote about it seriously. What he suggested is this: the cells and tissues of the body adapt to their environment during the life of an animal. Information from these cells is then shed into the bloodstream as "gemmules." These "gemmules" circulate in the blood and eventually lodge in the reproductive organs,

Gregor Mendel: Planting the Seeds of Genetics

where they become incorporated into the sex cells. This theory had similarities to the Lamarckian[2] idea that characteristics acquired during an organism's lifetime—directed changes in response to the environment—may be inherited by its offspring. The key difference was that Darwin's changes were undirected, providing a source of variation that could be inherited. Pangenesis and the notion of gemmules as the basis of inheritance was, of course, ultimately wrong.

The irony is that what Darwin really needed was Mendelian genetics. Mendel actually shows in his paper that if you start with a hybrid with all seven of the character pairs that he studied, the next generation will be composed of *one hundred and twenty-eight different combinations of those characters*. That's inherited variation for you, and no blending at all. But, of course, Darwin never read Mendel, and if he had it is very probable that he wouldn't have grasped the significance of the work. Anyway, his theory of pangenesis was published and discussed, and heavily criticized by many of his own admirers (see pp. 119ff.) for Francis Galton's experiment to test the theory), and remains as a testament to the fact that even Darwin stumbled occasionally.

In this lab in Tübingen in 1869, Swiss chemist Friedrich Miescher isolated a substance from the nuclei of dead white blood cells that he named nuclein—which would later be known as DNA.

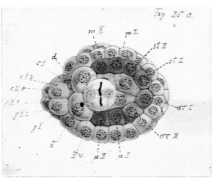
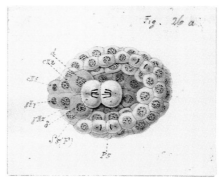

But if even Darwin missed Mendel's work, the question arises, how *did* biologists come to see Mendel as so important? The process took over thirty years and a gradual accumulation of concepts. In effect, biology had to grow into an appreciation of his significance.

The Nucleus

In 1866 Ernst von Haeckel speculated that the nucleus was the transmitter of inheritance. Shortly afterward (1869), the Swiss chemist Friedrich Miescher isolated a substance from the nuclei of dead white blood cells that he named nuclein. To get his nuclei, he used pus from bandages collected from the local hospital, which sounds revolting, but has a firm logical base—Miescher knew that the white blood cells in pus had large nuclei and little cytoplasm and would therefore be a good source of nuclear material. Later, when it was found that

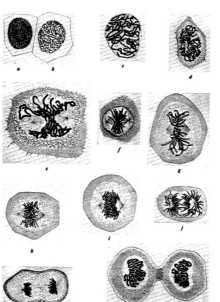

nuclein contained protein that could be digested away to leave a gooey, acidic gel behind, the term *nucleic acid* was invented. That was in 1889. By 1893 a pupil of Miescher's, Albrecht Kossel,[3] had performed a crude chemical analysis of nucleic acid and found it to contain phosphorus, sugar, and a mixture of four nitrogen-containing compounds that he identified as adenine, cytosine, guanine, and thymine. These are familiar to all with a basic knowledge of biology, often through their initial letters—A, C, G, and T. It was clear to some that this new substance must be important, but what could they do with it? They would have to wait half a century before that question could be answered.

Chromosomes and Fertilization

Meanwhile, in 1873 Anton Schneider had observed and recorded the appearance of chromosomes in dividing cells—correctly

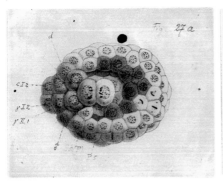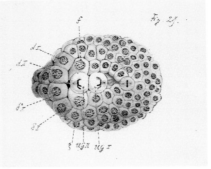

Drawings of cell division, by Theodore Boveri. Through countless investigations of horse nematode and sea urchin eggs, Boveri argued that individual chromosomes possessed different hereditary qualities.

noting that these filaments line up on the equator of the cell, double, and divide between the daughter cells. Although others—possibly including von Nägeli— had seen such filaments before, this was the first accurate observation of the process of mitosis. Two years later Oscar Hertwig, working with the sea urchin, observed that fertilization involved the fusion of the nuclei of sperm and egg to form a zygote—the first cell of the new organism. Later (1878) Hertwig made the discovery that the sperm and egg nuclei held exactly half the number of chromosomes of the zygote, and shortly after, using a microscope with the newly invented oil immersion lens and employing equally newly invented aniline dyes to stain tissues, Walther Flemming observed cell division in organisms as diverse as lilies and salamanders. In 1882 in his book *Zellsubstanz, Kern und Zelltheilung* he published beautifully accurate drawings of mitosis, which have influenced our understanding of cell division ever since. Roux (in 1883) remarked on the longitudinal splitting of these threads, so that each daughter cell received a copy, and in 1887 Weissmann made the significant speculation that in the formation of gametes (sperm and egg) a *reduction* division would be necessary in order to halve the chromosome number prior to fertilization, which would then restore the original number. Such a reduction division was discovered in the same year by Theodore Boveri and named *meiosis*. It wasn't until 1888 that Waldeyer introduced the term "chromosome," but it was already clear to some biologists that these threadlike structures might be connected with inheritance. By 1895 the American cytologist E. B. Wilson was writing: ". . . [this] seems to show that the chromosomal substance, the chromatin, is to be regarded as the physical basis of inheritance." He went on: "Now, chromatin is known to be closely similar to, if not identical with, a substance known as nuclein, which analysis shows to be a tolerably definite chemical compound composed of nucleic acid and albumin. And thus we reach the remarkable conclusion that inheritance may perhaps be affected by the physical transmission of a particular chemical compound from parent to offspring."

OPPOSITE: In 1882 the German cytologist Walther Flemming observed the process through which bundles of chromosomes split into pairs, evenly divided between the resulting two new cells. He called this process mitosis (from the Greek word for the thread in a warp, [*mitos*] plus *osis*, for "process"). Cell biologists of the early 1900s became increasingly convinced that this pattern of cell division reflected the process of heredity.

August Weissman was the first to propose a hereditary role for the chromosome, suggesting in 1893 that the hereditary units ("ids") were transmitted independently from each parent.

This is astonishingly percipient, although it is, of course, mere speculation. Another important speculator has already been mentioned: the German biologist August Weissmann. In 1892 in his book *Germ Plasm* he propounded his principle that inheritable material (*germ plasm*) is distinct from the body's material (*soma*). If true, this separation effectively removed the possibility that changes in body cells might influence the inherited information that is carried in the germ cells. One of the most important theoretical ideas in biology, Weismann's germ-line theory, effectively put an end to the Lamarckian speculations on the inheritance of acquired characteristics. From now on the reproductive cells would be seen as carriers of inheritance information and quite distinct from the somatic cells of the adult plant or animal. At last scientists were ready to come to terms with Mendel's findings. The stage was set for the rediscovery.

A Spat among Botanists

The usually quoted story is this: three botanists, independently of one another, rediscovered Mendel's work in the spring of 1900. They were Hugo de Vries, a Dutchman; Carl Correns, a German; and Erich von Tschermak, an Austrian. That's the story, but actually there is rather more to it than that.

That spring Hugo de Vries, Professor of Botany at the University of Amsterdam, published two papers. One, in March, was in French and came out in the *Comptes Rendus de l'Académie des Sciences* of Paris. The other, longer and more detailed, was published a few weeks later in the Proceedings of the German Botanical Society. Both were outlines of de Vries's own experimental work on plant inheritance, which he had carried out since 1892.

What de Vries had discovered was that if two plants differing in one character were hybridized then the hybrid would show either the male parent's character or the female's. However, the *other* character was still there, hidden in the hybrid, and would emerge in the next generation if the hybrids were mated together. He called the character that appeared in the hybrid, the *dominant*. The latent, hidden character he named the *recessive*. When the hybrids were bred together, the characters "segregated out" (de Vries used that phrase) into the pollen cells or the eggs cells, and the offspring of the hybrids showed a characteristic ratio of three dominant to every one recessive, a 3:1 ratio.

All this is, of course, exactly what Gregor Mendel had discovered in the 1860s, even as far as the use of the same terms—*dominant* and *recessive*—as Mendel had used. However, at some time during the preparation of these two papers an envelope arrived at de Vries's office from his fellow biologist, Professor Beijerinck of Delft University. "I know you are studying hybrids, so perhaps the enclosed offprint of 1865 by a certain Mendel that I happen to possess is still of some interest to you," Beijerinck had written. De Vries took out the enclosed academic paper and unfolded it on his desk. It was written in German, and was rather old-fashioned in appearance. Its title was *Versuche über Pflanzen-Hybriden*. Gregor Mendel had finally come to light.

De Vries read the *Versuche* through with growing amazement and horror: everything that he had discovered through his own plant-breeding program had already been done, thirty-five years earlier and in a much more systematic and thorough experimental approach than his own. Everything. And worse: his own *explanation*, the idea of paired units of inheritance being transmitted, one from the male parent and one from the female, into the offspring, and then separating out when the offspring make either pollen or eggs, had already been put forward. And all this by some unknown Austrian friar.

Austrian Erich von Tschermak was one of three scientists to "rediscover" Mendel's work in the spring of 1900.

By the time de Vries had finished his reading he must have been shattered. A similar thing had happened to Darwin in 1858, when he received a letter from Alfred Wallace, a botanist working in Malaysia. This letter detailed the theory of Natural Selection that Darwin himself had been working on but had never published. "All my originality, whatever it may amount to, will be smashed," were Darwin's words. It had taken the persuasion of his friends, who knew about his ideas, to convince him to publish a paper jointly with this Wallace rather than surrender the honor of the discovery to Wallace alone. In the case of Mendel's paper, de Vries did the decent thing as well: he turned back to the paper he was working on and revised it to include mention of Mendel, thereby surrendering priority to the obscure friar from Moravia.

This, as related by de Vries's assistant, T. J. Stomps, is how Mendel's work was rediscovered. However, there is a curious and conspicuous fact about those two papers that de Vries published in the spring of 1900. The German paper, the *second* one, may mention Mendel—three times to be precise, and once in a footnote—but the *first* paper, the brief summary published in French in the *Comptes Rendus* of the Academy of Science, has no mention of Mendel at all. The loyal Stomps claimed that the reason for this was that the French paper was merely an abbreviated version of the subsequent German version, and simply didn't include the Mendel parts, but this is patently untrue—the two papers are

One of Mendel's "rediscoverers," Hugo de Vries was a botanist at the University of Amsterdam who published the results of his own experiments on plant inheritance in 1900.

materially different. The reader can judge an extract. This is the close of de Vries's French paper:

> The totality of these experiments establishes *the law of segregation of hybrids*[4] and confirms the principles that I have expressed concerning the specific characters considered as being distinct units.

and this the German:

> From these and numerous other experiments I draw the conclusion that the law of segregation of hybrids as discovered by Mendel for peas finds very general application in the plant kingdom and that it has a basic significance for the study of the units of which the species character is composed.

It is difficult not to suspect that in the first extract he is claiming the law as his own discovery—and acknowledging Mendel's priority only reluctantly in the second.

Robert Olby, in his book *Origins of Mendelism,*[5] has tried to put a benevolent gloss on this, suggesting that the letter from Beijerinck came *after* the French paper had been sent off to Paris but *before* the German paper had been completed. But that doesn't allow for de Vries's own words in the footnote reference to Mendel in the German paper: "This important treatise is so seldom cited

Gregor Mendel: Planting the Seeds of Genetics

that I first learned of its existence after I had completed the majority of my experiments and had deduced from them the statements communicated in the text."

From this it doesn't seem as though the discovery of Mendel's work came during the preparation of de Vries's papers, but *during the actual experimental work* that led up to them. So why did Stomps put out this phoney story? Because phoney it surely is: elsewhere de Vries himself admits that he did *not* first hear of Mendel in 1900 but much earlier—in the 1890s, in fact. In a letter written to the American plant-breeder L. H. Bailey in 1905 or 1906 (the date is uncertain), de Vries acknowledges that he encountered Mendel's work through the bibliography accompanying Bailey's own paper of 1892, *Cross-Breeding and Hybridizing*. Bailey doesn't actually refer to Mendel in the text of this paper—this was yet another example of the bibliography of W. O. Focke's book (see pp. 72–73) being lifted wholesale—but de Vries says he "looked [the paper] up and studied it." That ties in with the footnote in his paper and puts de Vries's discovery of Mendel back into the 1890s. So why did he keep quiet about it until 1900 and the arrival of Beijerinck's letter on his desk?

One must not underestimate Hugo de Vries. He was both a brilliant speculative biologist and an excellent experimenter. In 1889 he had published a remarkable book called *Intracellular Pangenesis* in which it was clear that he already had presumed the idea of "units of inheritance" existing in cells and being passed on to their descendents via the nucleus. Curiously, in homage to Darwin, he called these units *pangenes*. De Vries's ideas weren't always correct, but *Intracellular Pangenesis* shows remarkable insights and is a fine demonstration of how biological thought had been changing in the years since Mendel had completed his work. It is certain that Mendel's own work on the garden pea would have meant a lot more to de Vries than it had to any of Mendel's direct contemporaries, for de Vries had already repeated Mendel's experiments, using a large number of different species. Nevertheless the question remains: why did de Vries keep quiet about Mendel when he first read the *Versuche* (probably) in 1895? And why wasn't Mendel even mentioned in the paper in the *Comptes Rendus de l'Académie des Sciences*? Was de Vries perhaps hoping that he could get away with not mentioning Mendel at all, and take the honor for the discovery himself? Did the appearance of the offprint on his desk, sent by Professor Beijerinck, force him to acknowledge the existence of Mendel's work?

If he did think like this, then almost immediately his fears were confirmed. On the morning of April 21, 1900 Carl Correns, of the University of Tübingen received, ironically direct from de Vries himself, an offprint of the note de Vries had sent to the *Comptes Rendus*—the one that makes no mention of Mendel.

One of Mendel's "rediscoverers," the German botanist Carl Correns also published his research on plant inheritance in 1900. His paper calls the Law of Segregation of Hybrids "Mendel's Law."

Correns was another botanist working in the area of hybridization and fertilization—in particular, he worked with *Pisum sativum*, the garden pea. He read de Vries's note through and immediately (*immediately*—he claimed that he completed it the next day!) wrote a long and indignant article to the German Botanical Society entitled *G. Mendel's Law Concerning the Behavior of Progeny of Varietal Hybrids*. It starts:

THE LATEST PUBLICATION OF HUGO DE VRIES: *Sur la loi de disjonction des hybrides*, which through the courtesy of the author reached me yesterday, prompts me to make the following statement: In my hybridization experiments with varieties of maize and peas, I have come to the same results as de Vries, who experimented with varieties of many different kinds of plants, among them two varieties of maize. When I discovered the regularity of the phenomena, and the explanation thereof—to which I shall return presently—the same thing happened to me which now seems to be happening to de Vries: I thought I had found *something new. But then I convinced myself that the Abbot Gregor Mendel . . . had, during the sixties, not only obtained the same result through extensive experiments with peas . . . but had also given exactly the same explanation, as far as was possible in 1866.*[6]

So Carl Correns had *also* known of Mendel's work and had kept quiet about it! In an attempt to forestall criticism he adds: "At the time (of reading Mendel's paper) I did not consider it necessary to establish my priority for this 'rediscovery' by a preliminary note, but rather decided to continue the experiments further."

Later in the paper de Vries's use of the terms *dominant* and *recessive*—certainly a bit suspicious—provokes Correns to fury: "Mendel named them in this way, and, by a strange coincidence, de Vries now does likewise." That sarcasm— *by a strange coincidence*—burns through the page. Clearly Correns had no doubt. He didn't suggest that the good professor might not have been aware that the obscure Gregor Mendel got there first: clearly he was certain that de Vries knew about Mendel. How? One can't help wondering about the origin of the *Versuche* offprint—"which I happen to possess"—that Beijerinck passed on to de Vries. Did it perhaps come from Correns? Because, by a strange coincidence, Correns had been a pupil of Carl von Nägeli and was a close friend of that man's family. Indeed, he married von Nägeli's great-niece. Through this close contact he also had access to von Nägeli's papers and later he was to edit the letters that Gregor Mendel had written to von Nägeli. Of course also there among von Nägeli's papers was the offprint of the *Versuche* that Mendel had sent on New Year's Eve 1866.

Hugo de Vries had the last laugh in this unpleasant academic spat. His *German* paper—the one that gave due recognition to Mendel—was published in the Proceedings of the German Botanical Society at exactly the time that Correns was checking the proofs of his own paper attacking de Vries. How galling it must have been for Correns to discover in that paper the very mention of Mendel that he had criticized de Vries for omitting in the French version. It was, apparently, too late to revise his paper and he was forced to add a postscript:

> In the meantime de Vries has published in these Proceedings (No. 3 of this year) some more details concerning his experiments. There he refers to Mendel's investigations, which *were not even mentioned in the Comptes Rendus.*[7]

My own postscript is this: Correns clearly was about to complete his own article when he received de Vries's French paper—he couldn't possibly have completed it in the single day that he suggests. Was he also planning to keep quiet about Mendel? Did he do a hasty and self-righteous revision in order to occupy the moral high ground? We will never know. All that is certain is that out of this academic spat came a law. De Vries had called it the Law of Segregation of Hybrids, but Correns clearly and unequivocally ascribed it to G. Mendel—"Mendel's Law." Thereby begins the metamorphosis of Mendel the man into Mendel the patron saint of genetics.

The role that the third man, von Tschermak, played in all this is easily described. He also read de Vries's French paper and hurriedly came out with the results of his own work, also with the garden pea. But his paper gives no real indication that he had explained his work to the degree that de Vries and Correns had, and in his postscript (there is only one other mention of Mendel in the body of the text) he seems rather relieved that the rediscovery of Mendel has effectively stopped any argument about priority among the three discoverers. "The simultaneous 'discovery' of Mendel by Correns, de Vries, and myself appears to me especially gratifying," he writes.

Mendel's Disciple

A saint needs a disciple and Mendel soon acquired his. It was not de Vries, who in 1910 pointedly refused to contribute to a Mendel memorial in Brünn, and nor was it Correns or von Tschermak. Instead it was an Englishman called William Bateson.

British zoologist William Bateson (right), an enthusiastic promoter of "Mendelism," gave the study of inheritance a new name—"genetics." R. C. Punnett (left) devised the concept of the "Punnett Square," well known to all introductory biology students as a way to depict inheritance ratios.

Two journeys feature in the mythology of genetics, each at either end of the twentieth century. One is the night-time drive that Kary Mullis was taking in 1983 on Route 101 (or was it Highway 128?—like all good myths there are different versions) in California when he dreamt up the idea of the Polymerase Chain Reaction (see p. 146); the other was the train journey made from Cambridge to London by William Bateson in early 1900. It is during this journey that he is reputed to have read Mendel's paper, and as a result revised his planned talk to the Royal Horticultural Society that was to take place at the end of the journey. Instead of explaining de Vries's work he announced Mendel to the world.

Whatever the truth of this story, there is no doubt that Bateson *did* read the *Versuche* in 1900 (probably after reading de Vries's paper), grasped its importance, and set about spreading the word. Bateson became a true disciple, unbending in his loyalty even though occasionally misguided in his conclusions. It was Bateson who first had the *Versuche* translated into English and published; it was he who first traveled to Brno to discover the roots of Mendel's ideas; it was he who first applied Mendelian principles to animals (particularly poultry); and it was he who both clarified what Mendel had discovered and found the first exceptions to what were by then being called Mendel's Laws.

Mendel's Laws

As formulated in the early twentieth century, these are the familiar principles that are found in elementary biology textbooks right up to today. The Law of Segregation states that genes come in pairs, that each pair controls a single inherited character, and that when an adult organism makes gametes (sperm cells or egg cells) the genes of each pair *segregate* into the gametes, so that each gamete receives only one gene from each pair. Thus for every gamete containing one of the two genes, there is a second gamete containing the other. The two genes of a pair came to be known as *allelomorphs*, a term that we owe to Bateson and which is now always shortened to *allele*. In a normal cell the alleles may be identical—**AA** or **aa**—or they may be different from each other—**Aa**. The law of segregation is the embodiment of the crucial idea that genes are distinct *things*—particles or units of inheritance; what in 1909 Wilhelm Johannsen first called "genes." In 1902 Bateson had coined the term *homozygous* to describe an individual with both alleles the same (**AA** or **aa**); and *heterozygous* for an individual in which the two alleles were different (**Aa**). At the same time as he invented gene, Johannsen also coined *genotype* (the genetic composition of an organism) and *phenotype* (the outward appearance of the organism). Genetics was beginning to assemble its vocabulary.

The Law of Independent Assortment, the second of Mendel's Laws, takes the Law of Segregation one step further. Independent assortment is the principle that any pair of genes will segregate (i.e., be inherited) independently of any other pair. If an organism is **AaBb** there is an exactly equal chance of it producing the following combinations when it makes gametes: **AB**, **Ab**, **aB**, **ab**—which example not only uses Mendel's notation but looks as though it comes directly out of his famous paper.

It is worth noting that although these two "laws" are derived from Mendel's work and, to the modern reader, implied by what he wrote in the *Versuche*, they are nowhere expressed in quite such unequivocal terms by Mendel himself. And whereas the concept of segregation has stood the test of time, almost immediately the principle of independent assortment was shown to have exceptions, occasions when it did not seem to hold true. In other words, examples were soon found when **A** or **a** were *not* inherited independently of **B** and **b**. This would puzzle Bateson and inspire the Americans Walter Sutton and Thomas Hunt Morgan to further momentous discoveries. It was the beginning of modern genetics.

PLATE 5

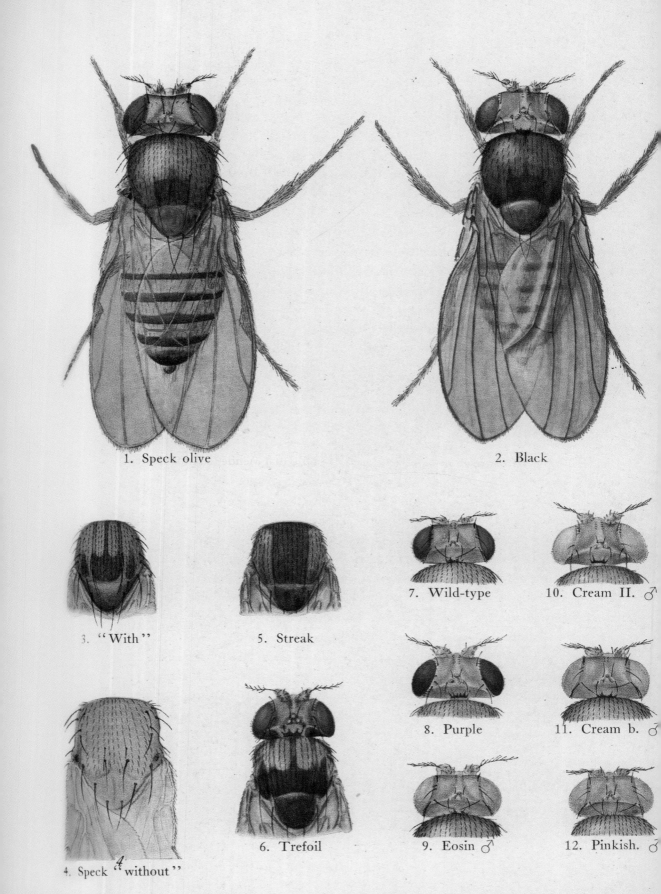

1. Speck olive

2. Black

3. "With"

5. Streak

7. Wild-type

10. Cream II. ♂

8. Purple

11. Cream b. ♂

9. Eosin ♂

12. Pinkish. ♂

4. Speck "without"

6. Trefoil

The Mendel Legacy

After 1900 many biologists jumped onto the Mendelian band wagon, determined to find out whether Mendel's "laws" held across the whole spectrum of living organisms. Ironically, it fell to Bateson himself, in 1905, to stumble onto the first case where the so-called Law of Independent Assortment did *not* hold.

Chromosomes

Working with his assistants E. R. Saunders and R. C. Punnett, Bateson was investigating the inheritance of flower color (purple vs. red) and pollen grain shape (long vs. round) in the sweet pea (*Lathyrus odoratus*). Purple flowers and long pollen grains, they found, are the two dominant phenotypes. They had obtained first generation (F_1) heterozygotes (**PpLl**) from pure-breeding parents (**PPLL** x **ppll**) and just as they expected, when they self-pollinated these double heterozygotes they found that they *did* get the expected four different combinations of phenotype: purple and long, purple and round, red and long, red and round. However these did *not* appear in the 9:3:3:1 ratio that Mendelian independent assortment predicted if the genes for these characters had been passed on to the pollen and ovules independently of each other. Instead, they found these results:

F2 results from crossing PpLl x PpLl				
Phenotypes	Found	%*	expected	ratio
Purple, Long	284	75	214	9
Purple, round	21	6	71	3
red, Long	21	6	71	3
red, round	55	14	24	1
total	381	*not equal to 100% because of rounding errors		

The ten-day life cycle of the fruit fly (*Drosophila melanogaster*) made it an ideal research subject for geneticists in the early 1900s.

It is clear that there are far too many purple-flowered plants with long pollen grains, and red-flowered plants with round grains, i.e., having the combinations of the *grandparents*, and

Studying the chromosomes of the grasshopper, Walter Sutton surmised that the pairing of maternal and paternal chromosomes that occurs during cell division constitutes "the physical basis of the Mendelian law of heredity." During the process of mitosis chromosomes separate (fig. 1) and align around the midplate of the cell and attach to spindles (fig. 2). The spindles then pull half of the chromosome pairs to either end of the cell (fig. 3), where they collect and wait for the cell to be divided.

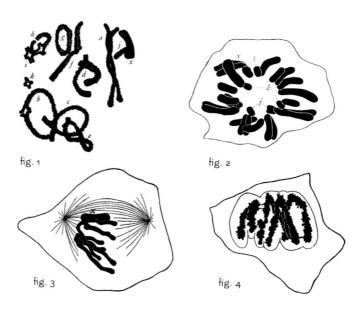

fig. 1

fig. 2

fig. 3

fig. 4

OPPOSITE: Variations of the sweet pea flower, from *Mendel's Principles of Heredity*, by William Bateson, 1909. The sweet pea flower (*Lathyrus odoratus*) was the first test subject that did not subscribe to Mendel's Law of Independent assortment, indicating that some traits are linked.

too few of the new combinations, purple flowers with round pollen, or red flowers with long pollen grains. This result was so far away from the expected as to be completely bewildering. And Bateson and Punnett certainly were bewildered. They called it "coupling"—the two dominant genes and the two recessive genes seemed to be coupled together—but they couldn't explain why it occurred.

What the two Englishmen didn't do—in fact they deliberately took a position against it, which seems bizarre to us nowadays when the solution seems so obvious—was to follow the work of two cell biologists who had published important observations just a couple of years earlier. One of them, Theodore Boveri, was schooled in the nineteenth-century tradition of German science; the other, Walter Sutton, was a Kansas farm boy. Their pairing in the history of genetics is perhaps symbolic of the change that was taking place as one century gave way to the other—the balance of genetic discovery and understanding was shifting across the Atlantic, away from Europe toward the United States.

Walter Sutton worked on an animal that was familiar to him from childhood—the grasshopper, *Brachystola*. He watched (and drew beautifully) the behavior of chromosomes in meiosis (gamete production) and confirmed Boveri's observation that the number of chromosomes was reduced to half when sperm and egg cells were made. He also noted the pairing of maternal and paternal chromosomes at the start of the process and their separation into the gametes. Most significantly, he concluded his paper with this remark:

Gregor Mendel: Planting the Seeds of Genetics

Plate V

Parent
White: flat standard

F₁

Parent
White: hooded standard

1

3

2

F₂ light wings

4

5

6

F₂ dark wings

7

8

9

F₂ whites

10

11

1. Emily Henderson.   2. Blanche Burpee.   3. Purple Invincible, F₁.   4—11. The various F₂ types obtained by self-fertilising F₁.   4. Purple Invincible.   5. Duke of Westminster.   6. Painted Lady. 7—9. Corresponding dark winged types.   7. Purple, with purple wings.   8. Duke of Sutherland. 9. Miss Hunt.   10 and 11. F₂ whites.   Notice that there is no *hooded* red.

Nettie Stevens investigated the chromosomes of the fruit fly from her lab at Bryn Mawr College. Stevens was one of the first American women to gain international scientific recognition.

I may finally call attention to the probability that the association of paternal and maternal chromosomes in pairs and their subsequent separation during the reducing division as indicated above may constitute the physical basis of the Mendelian law of heredity.

As an example of a knowing understatement this surely ranks alongside Watson and Crick's famous aside in their 1953 letter to *Nature* (see p. 139). What Sutton implied, of course, is that chromosomes are the "things" that are inherited, and that genes (the effects of which the breeder is watching) must therefore be *on* the chromosomes. Obviously, if two different genes are close together on the same chromosome they will tend to stick together, just as Bateson had found with flower color and pollen shape in sweet pea. What Sutton's proposition would *not* do, however, is explain how it is possible for two such genes to become "uncoupled" as they appeared to do in 12 percent of the occasions in Bateson's 1906 experiment.

Had Bateson pursued the chromosome line of thought he might today be remembered as more than merely Mendel's great supporter—his status as one of the true greats of genetics would probably be much more elevated. But he shunned chromosomes with a fervor that was almost religious and instead it fell to a team of Americans, like Sutton working at Columbia University, to explain what was going on.

Fruit Flies

Insects were fast becoming the most significant group of organisms in cytology and genetics. In 1902 C. E. McClung—Sutton's first supervisor at the University of Kansas—had hypothesized that one of the chromosomes in insects (earlier named "X" by German cytologist Hermann Henking because it was enigmatic) was responsible for determining sex. This was fully confirmed in 1905 by the celebrated American cytologist E. B. Wilson and, independently at Bryn Mawr, by Nettie Stevens, one of the first American women scientists to achieve international fame (and surely an inspiration to Barbara McClintock (see pp. 148–49). They found that the females of the various species they investigated always contributed an X chromosome to each offspring, whereas the males contributed *either* an X *or* a Y chromosome (or, in some species, no sex chromosome at all). Getting two X chromosomes actually made the offspring female, while an X and a Y (or, in some species, an X and a *missing* sex chromosome, XO) made the animal male.

This was the moment for the common fruit fly, *Drosophila melanogaster* to take its place at center stage of genetics. So successful has *Drosophila* been as an experimental organism that it still occupies a major place in research almost exactly a century after its introduction in 1908, when a fellow biologist suggested to the outspoken Professor of Experimental Zoology at Columbia, Thomas Hunt Morgan, that he might consider using the fly in his work. *Drosophila* was, and is, ideal. Small (about three millimeters long) and annoying, it is a pest attracted to over-ripe fruit and has a tendency to go swimming in your glass of wine on an autumn evening. But it loves bananas (one of the first media used to feed *Drosophila* in the lab), it can be reared easily in quarter-pint milk bottles with a cotton wool plug in the opening, and it completes its life cycle in under two weeks. This means that *Drosophila* can go through about twenty-five generations in the same time that most plants go through a single generation—with *Drosophila* there would be no more waiting for next spring to see the results of a cross. Not only that, but the flies are easily sexed, can be readily anaesthetized, and have a startling reproductive rate. And virgin females (essential for ensuring that a female has not already been impregnated by one of her brothers) can be identified using a magnifying lens and a practiced eye.

Altogether the fruit fly seemed ideal. The only trouble was that they all looked the same—the so-called wild type. Morgan was particularly interested in de Vries's theory of mutation and he began by breeding the flies under varied conditions to see whether mutations (sudden, inheritable changes in appearance) would appear. For *two years* he bred the flies before he had any success, a degree of tenacity to match Gregor Mendel's. One of the earliest, and the first dramatic mutation, that he found concerned eye color. The wild type fly has brilliant red eyes, but one day Morgan found himself looking through the walls of a milk bottle at a male fly with white eyes.

Sex Linkage

One imagines the agonizing care with which that first, historic white-eyed fly was removed from the company of its siblings. Having been isolated it was then mated with a selected "wild

Support for the chromosome theory came in the early 1900s. Thomas Hunt Morgan looked at external characters—such as eye color and wing size—over generations of fruit flies.

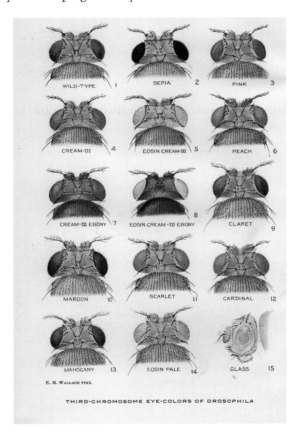

THIRD-CHROMOSOME EYE-COLORS OF DROSOPHILA

A bottle of fruit flies used in the Columbia University Fly Room

type" female (red eyes). Morgan might have expected that all the offspring would be red-eyed—and indeed they were. So far, so Mendelian. White eye was clearly recessive to red eye. And when he crossed the males with the female hybrids, he obtained the expected three red-eyed for every one white eyed in the offspring, exactly as Mendel would have expected. Except that *all the white-eyed flies were male.* It was only when he crossed the original white-eyed male with one of his red-eyed daughters (an advantage of the short life cycle of *Drosophila!*) that he obtained some white-eyed females.

What seemed to be happening—Morgan's suggestion in his paper[1]—was that the gene for white eye was being inherited *along with the X chromosome.*

It is instructive to read Morgan's original paper on this work and compare his manner of expression with Mendel's in the *Versuche*. Firstly, Morgan's symbolic notation is much harder to follow than Mendel's and at times it is actually wrong. Secondly, Morgan never actually makes a precise statement such as "therefore the white-eye gene must be *on* the X chromosome." Ironically, if Morgan were to be judged by this paper alone we would even now be debating whether he really understood what he had discovered—exactly as we do with Mendel. However, Morgan was the supreme collaborative worker. There was a vibrant exchange of ideas and theories within the Fly Room at Columbia and over the next decades, he and his graduate students published a whole succession of papers that show how their understanding of chromosomes and genes developed. Mendel, of course, had no one of his own intellectual stature to assist him in his work or to stimulate him to develop his ideas.

Within a short time the team in the Fly Room had discovered other examples of mutations that were always inherited with the X chromosome, for example, *vermilion eye color* **p** and *rudimentary wing* **m**.[2] All such genes were inherited together, since they were located on the same chromosome and were, in fact, linked. This goes right against Mendel's so-called second Law of Independent Assortment, and is exactly what Bateson and Punnett had already discovered in sweet pea with flower color and pollen grain shape. Thus, in *Drosophila*, if a hybrid female fly is produced from a wild type mother and a vermilion eyed, rudimentary winged father, she will have the following gene composition, each pair of genes sitting on the animal's X chromosome like this:

$$\frac{\text{P M}}{\text{p m}}$$

Gregor Mendel: Planting the Seeds of Genetics

LEFT AND BELOW: The Fly
Room at Columbia
University

The Mendel Legacy

When this female produces eggs, they will either contain the X chromosome with *both* the dominants (**P** and **M**), or the *other* X chromosome that carries both the recessives (**p** and **m**). So her eggs would either have:

1) **P M**

or

2) **p m**

Which, judging by the result of breeding, is exactly what happened—except that in a few cases the other combinations, **P** with **m** and **p** with **M**, *were* found. In these cases it seemed that the dominant **P** gene had somehow "crossed over" to the other chromosome and ended up attached to the recessive **m** gene, and, presumably at the same time, the **p** gene had crossed over the other way and become associated with the **M**. On these occasions crossing over had effectively broken the linkage of the two genes, achieving what Morgan's group came to call "recombinants":

3) **P m**

and

4) **p M**

The different pairs of genes on the X chromosome seemed to show this "recombination" to different degrees. Between some pairs it was a rare event; between others it seemed to be relatively common—i.e., some genes seemed to be more strongly linked together than others. In the case quoted here, recombinations (3 and 4 above) occurred in 109 offspring out of a total of 458—that is, in 23.8 percent of occasions. Morgan suggested that this strength of association, this "recombination frequency," was some kind of indication of how far apart the genes were on the chromosome: the stronger the association, the closer the genes are.

This is where one of Morgan's co-workers named Arthur Sturtevant made his revolutionary proposal[3]: knowing the recombination frequencies of the X-linked genes, he realized that it should be possible to construct a *map* showing their relative positions and their relative distances apart on the chromosome. In the example given, **P** and **M** would be 23.8 units apart. **P** and **R** (another gene affecting wing form) showed 3 percent recombination and are therefore 3 units apart. The recombination frequency between **R** and **M** ought to show whether **R** lies on the **M** side of **P** or on the other side. Fired up by his idea and working through the night, Sturtevant created a map for five known genes on the X chromosome.[4]

Gregor Mendel: Planting the Seeds of Genetics

```
B   C                P        R                              M
|   |                |        |                              |
0.0  1.0            30.7    33.7                           57.6
```

Gene Map for X Chromosome of *Drosophila* after Sturtevant, 1913

Drosophila notation underwent changes. In this map, taken from Sturtevant's paper, **B** is the gene for yellow **B**ody color (it is now called **y**); **C** is the original white-eyed **C**olor, nowadays known as **w**; **P** was **P**ink eye, later renamed vermilion, **v**; and the **R** and **M** refer to rudimentary and miniature wing, which Morgan got wonderfully muddled by making **R** mean "not **R**udimentary" and **M** mean not "not **M**iniature."

The publication of this first-ever chromosome map heralded a revolution to rank alongside the rediscovery of Mendel's paper and the elucidation of the genetic code by Nirenberg and others in the 1960s (see pp. 142ff.). Virtually all gene mapping since Sturtevant's paper, right up to and including the Human Genome Project, is based on this fundamental principle, that the degree of recombination between any two genes on the same chromosome gives a measure of their physical separation. And fittingly the basic unit of gene mapping is named after Morgan himself—the centiMorgan is the distance between two genes that show 1 percent recombination between them.

It is hard to believe, but Sturtevant was a nineteen-year-old undergraduate when he created this first gene map. He later confessed that he did the overnight work at the expense of an undergraduate essay! It is surely a testament to the openness of Morgan's laboratory and the manner in which he himself worked that such a young and inexperienced researcher could take all the credit for this. Indeed, it is almost impossible to overestimate the importance of Morgan and his work. Within a short time after the publication of Sturtevant's paper, the Fly Room team had moved onto the other chromosomes of *Drosophila* (there are only four in total), and gene mapping proceeded apace. Now, suddenly and startlingly, Mendel's hypothetical *Elemente*, were being given a precise location on bodies that are visible in the nuclei of cells. Indeed, genes appeared to be organized on the chromosome like beads on a string.

Young Arthur Sturtevant—a co-worker of Thomas Hunt Morgan's and just nineteen at the time—was the first to document sex linkage of chromosomes on a linear gene map.

Darwin and Mendel

The history of genetics is not a simple narrative leading from ignorance to knowledge. Science just isn't like that. There are

The work of Thomas Hunt Morgan and his colleagues led to the correlation of characters with individual chromosomes and eventually to "mapping" genes on the chromosome itself. Still, two decades and millions of fruit flies later, Morgan stated in his 1934 Nobel Prize lecture that geneticists were divided as to whether genes "are real or purely fictitious."

parallel stories, diversions and dead ends, sub-plots and reversals, threads as intertwined and convoluted as a protein molecule. At this time, while Morgan and his group were beginning their ground-breaking work, there was, for example, a great conflict between the Mendelians and the Darwinians. Indeed, Morgan himself was involved in the argument. The point is that when Mendel's work was rediscovered, it was considered by many not so much as a theory of inheritance but rather as an alternative mechanism for evolution. This was de Vries's interest: he was dissatisfied with Darwinian gradualism and wanted to find evidence for his concept of evolution proceeding by major jumps or "saltations." His own theory of mutations tried to account for this and at first Mendelism seemed to provide support. And Bateson, the man who most championed Mendel, was himself convinced that Mendelism offered a rival to what he considered the erroneous concept of Darwinian natural selection. Others climbed aboard the bandwagon and it wasn't until the 1920s and 1930s, when Morgan's discoveries about chromosomal inheritance had been fully digested, that biologists produced a synthesis of the two theories. Mendelian genetics, biologists such as J. B. S. Haldane and Sewell Wright pointed out, was just what natural selection needed. In fact it was the piece of the jigsaw that Darwin had found missing—a theory that could account for the production, generation after generation, of stable, inheritable variation on which natural selection could work. Ironically, Mendel had made this very point in the second part of his own famous paper—but no one had taken any notice.

These kinds of discussions and arguments were going on in the background, while others were making what we now know to be major strides forward and asking the next questions. In particular, what are genes made of, and what exactly do they do? Yes, you could see that they caused effects such as dwarf plants or white-eyed flies. But how did they achieve this? What was the chemistry behind it? Because by now the science of living things had staked out its territory clearly enough: life was a matter of chemistry. Complex chemistry, yes; but chemistry none the less. There was no mystery about it and no magic. It could be explained by understanding the behavior of molecules.

Inborn Errors of Metabolism

As early as 1902 a truly remarkable discovery was made by a London physician, Archibald Garrod. His report in the British medical journal *The Lancet,* entitled "The Incidence of Alkaptonuria: A Study in Chemical Individuality," was star-

Gregor Mendel: Planting the Seeds of Genetics

tlingly ahead of its time. Garrod worked with patients with the condition known as alkaptonuria, in which the sufferers, among other problems later in life, excrete a large amount of a chemical called homogentisic acid in their urine. This substance turns black on exposure to air and one of the early symptoms is therefore the staining black of a baby's diapers, so it is not the kind of thing a mother could miss. What Garrod understood was that alkaptonuria seemed not to be a disease in the normal sense of the word—a sickness that you caught from someone—but was what he later (1908) called an "inborn error of metabolism," an unusual and inherited inability to carry out a part of ordinary metabolism. In the case of alkaptonuria patients appeared not to be able to break down the amino acid tyrosine, which is a normal component of the diet, and this led to the build-up of an otherwise perfectly normal metabolic intermediate— homogentisic acid. This non-lethal condition has a variety of effects, including discoloration of cartilage (becoming visible in the nose and ears), arthritis, heart valve problems, and the discoloration of the urine. Garrod argued that this condition was an example of a rare recessive character inherited in *Mendelian* fashion (yes, he was already writing that in 1902!). He also suggested several other candidates for such an explanation—albinism was one—and by 1909 he had produced a book on the subject. For the first time a biochemical explanation for gene action was being proposed—a gene somehow confers (or fails to confer if it is defective) the ability to perform a particular step in metabolism. As Garrod himself expressed it:

> . . . it will be seen that in the case of each of [the several known inborn errors of metabolism] the most probable cause is the congenital lack of some particular enzyme, in the absence of which a step is missed, and some normal metabolic change fails to be brought about.[5]

One Gene: One Enzyme

By the 1930s, George Beadle and Edward Tatum were trying to investigate this idea of enzyme deficiency on an experimental organism that seemed far away from Garrod's patients at St. Bartholomew's Hospital in London—the mold *Neurospora crassa*. Using X-rays they developed a series of mutant strains of the mold that were each deficient in a single different metabolic step. It became clear from their work that Garrod's speculation was something of a law—each step in a metabolic pathway was controlled by a specific enzyme molecule, and each enzyme came from a specific gene that could be changed by X-rays—one gene: one enzyme. Enzymes were made of protein, so this soon became one gene: one protein.

Nucleic Acid

In 1922, Herman Muller, another of Morgan's pupils and the man who showed that X-rays could cause genetic mutation, wrote prophetically: "Perhaps we may be able to grind genes in a mortar and cook them in a beaker after all."[6] But despite knowing now more or less what genes *did*, still no one was any nearer to knowing what chemical they were made of—was it protein or was it Miescher's "nuclein"—the two substances that seemed to be the main components of nuclei? The searching and the arguing would take many years yet, largely because the techniques for visualizing molecular structures were still in their infancy. William Bragg and his son Lawrence had begun to use X-ray diffraction to elucidate the relatively simple and ordered structure of inorganic crystals, but at the time of their joint Nobel Prize for physics in 1915 and for some time after no one could have imagined that such techniques might one day be employed to work out the structure of complex biological molecules such as protein or nuclein.

However advances could be made in working out the composition of both types of molecule. Proteins were already known to be built up of amino acids in a long chain (Franz Hofmeister and Emil Fischer had shown this in 1902) and there were many different amino acids (at the time it wasn't certain how many), which could result in an almost infinite variety of proteins. But what was

By the 1930s, George Beadle (pictured here, with his assistant) and his partner, Edward Tatum, discovered that each step in the metabolic process develops from one specific gene.

nuclein? At the Rockefeller Institute in New York, the Russian-born Phoebus Levene, who had worked with Kossel (see page 89) on nucleic acid at the beginning of the century, improved the analysis of this curious, glutinous substance. It was to him that we owe the discovery of the sugar called deoxyribose and the linkage of these sugars together through phosphate groups. Also involved in the molecule were the four bases that Kossel had discovered, adenine, guanine, cytosine, and thymine. Bonded together—*phosphate-sugar-base*—these formed what Levene now called a nucleotide. And he also found that the two sorts of nucleic acid differed in their sugar component—ribose in one and deoxyribose in the other: hence, ribonucleic acid (RNA) and deoxyribonucleic acid (DNA).

Edward Tatum in his laboratory

So far, so good. However, it was here that Levene made his mistake: he decided that in nucleic acid the four different nucleotides were present in equal amounts and were probably bonded together in repeating units of the four different bases that he called tetranucleotides. Thus a single tetranucleotide would be G-A-C-T (that is, Guanine-Adenine-Cytosine-Thymine). Exactly how many tetranucleotides were joined together he wasn't sure ("the tetranucleotide theory is the minimum molecular weight and the nucleic acid may well be a multiple of it"[7]), but if the whole molecule contained nothing more than G-A-C-T repeats it was certainly rather uninteresting. Highly variable protein seemed a much better candidate for the real substance of the gene.

This tetranucleotide theory—Erwin Chargaff later referred to it as a "wet blanket of a hypothesis"[8]—acted for years as a damper on further investigation of nucleic acid structure. It led, for example, to the cytologist E. B. Wilson's dismissal of nucleic acid as a possible genetic material:

> . . . nucleic acids are on the whole remarkably uniform . . . In this respect they show a remarkable contrast to the proteins, which, whether simple or compound seem to be of inexhaustible variety. It has been suggested accordingly, that the differences between different "chromatins" [the granular material seen in nuclei] depend upon their . . . protein components and *not* upon their nucleic acids.[9]

But Levene's ideas were quite wrong, and ironically, the evidence against them lay within his own institution.

Nucleic Acid — Bond Distances and Angles

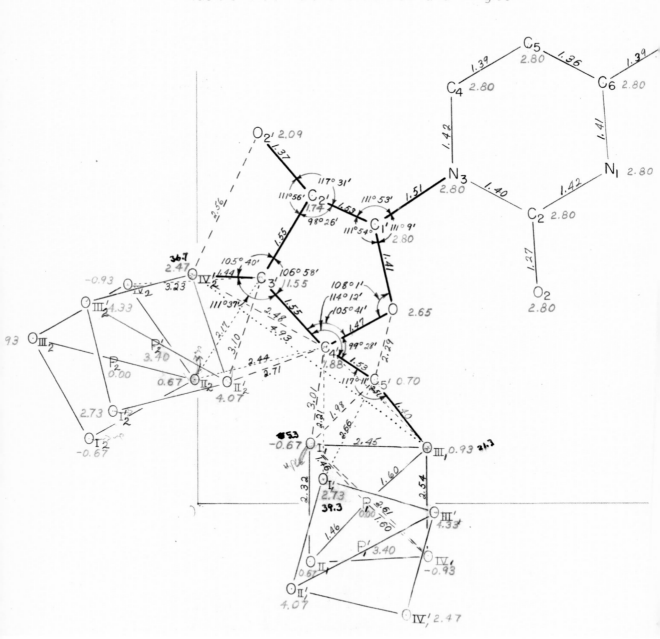

New Currents in Genetics

Genetics wasn't only something that happened in research laboratories. It also had a social impact. Even as Mendelism and Darwinian natural selection were being put together to give what came to be known as neo-Darwinism, a mutation of a non-genetic type arose.

Eugenics—Genetics' Ugly Sister

Eugenics (it means "well-born") grew out of the ideas of Francis Galton, a direct contemporary of Gregor Mendel, and came to a flowering in the 1920s and 1930s, just as Morgan and others were plotting the position of genes on chromosomes and Beadle and Tatum were showing that genes were responsible for making enzymes.

Unlike Mendel, Galton was born into a rich and privileged family—in fact the same family as Charles Darwin (they were cousins). Like Darwin, Galton did not show much promise in his early years. Originally destined for medicine, he soon gave up medical school in favor of travel in Europe, before returning to England and taking up a place at Cambridge to study mathematics. Here he suffered a nervous breakdown of some kind and left without taking a degree. Later he resumed his medical training, but when his father died leaving a sufficient fortune to keep his son comfortably off for the rest of his life, Francis abandoned medicine yet again. Now he was able to pursue whatever he fancied, and what he fancied turned out to be a strange variety of things: more travel, an obsession with measuring people, a penchant for mathematical analysis, and a conviction that he was a superior kind of person and that this superiority was inherited. At times he seems more than a little eccentric; certainly a man who, apparently in all seriousness, can write a "scientific" paper on the efficacy of prayer is rather unusual.[1]

Galton's role in the history of genetics is as curious as the man himself. Convinced by his cousin Charles's theory of evolution by natural selection, he was much less persuaded by the theory of pangenesis—indeed, he performed a series of experiments on rabbits, transfusing blood from one to another in order to disprove the theory that hereditary material was being carried in

This 1952 sketch of the structure of nucleic acid was made by Linus Pauling, a protein chemist in California who was also competing to determine the structure of DNA.

In the 1880s Francis Galton founded "Eugenics," a branch of study devoted to the "improvement" of the human race. Galton's idea of eugenics involved encouraging the healthiest and "fittest" people in a society to breed more.

their blood. Having succeeded in disproving this theory to his own satisfaction, he then set out to derive his own laws of inheritance. In his writings in the 1870s, he does give a quite clear idea that inheritance is "particulate," but most of his ideas were based on continuous variations rather than discontinuous variation in the Mendel tradition. Partly based on some experimental evidence derived from breeding sweet peas, Galton's first "law" was the observation that offspring of exceptional parents tend to "revert to the mean"—that is, they tend to be less extreme in nature than their parents. Galton applied this "law" to characteristics as diverse as stature and intelligence and although it represents a blending theory of inheritance under a different guise, his musings were not without merit: statistically (he was one of the founders of the science of statistics) the idea is quite sound. However, statistics is a means of description rather than a mechanism and Galton completely failed to provide any mechanism for inheritance.

Like Galton's first idea, his second "law of inheritance," has a certain plausibility. Children have one-half of each parent's "heritage," and therefore, on average, one-quarter of each grandparent's, one eighth of each great-grandparent's, and so on. In truth, this isn't really saying very much as long as one assumes (and this had become a general assumption by the 1880s, when Galton published the idea) that both sexes contribute equally to their offspring. Galton managed to obfuscate this truism with such a cloud of mathematics that it seemed to take on a profundity it didn't, and doesn't, deserve. However, in his constant reference to the inheritance of human attributes, particularly intellectual ability, he gave name to an idea that had long-lasting and tragic effects: eugenics. What had been a valid scientific investigation of human traits began to take on a political life of its own.

It was clear to Galton that humans possess inheritable characters just as plants and animals do, and that some of these characters must be desirable and some undesirable. It was his argument that it should be possible to have the "better" genetic types to breed and compel the "poorer" types to refrain from breeding. Thus the genetic stock would improve over time just as beef cattle or domestic dogs had been improved over the years. Encouragement of the "good" came to be known as positive eugenics; discouraging the "bad" was negative eugenics.

It is important, although difficult, to try and see this movement from the perspective of the times in which it was born. There was a genuine fear that the upper and middle classes—needless to say, considered hard-working and intelligent—were having fewer children[2] than the lower classes, who were, needless to say, perceived as feckless and stupid. Because of these different birth rates, the thinking went, the beneficial characteristics of the upper classes were being lost and the overall genetic quality of the race was threatened. Of course "the race" in question was invariably white and northern European. In the nineteenth century and into the twentieth, it was quite acceptable to consider that all other "races" were genetically—and therefore morally, intellectually, and physically—inferior.

In 1907, inspired by Galton's ideas, the Eugenics Education Society was founded in Britain with the explicit aim of spreading the doctrine of genetic improvement throughout the land. Galton himself became its honorary president in 1908.

Before this, Galton's protégé, Karl Pearson, a statistician of real originality, had developed his mentor's ideas of human measurement and formed the Biometric Laboratory at University College, London. Biometrics now became muddled up with the eugenic movement. If Galton was an enthusiast, Pearson was a fanatic—a cold, calculating measurer of man who claimed to be a socialist, but loathed the working class. His journal *Biometrika* became hugely influential, particularly in the United States, and in 1911 he became the first Galton Professor of Eugenics at London University, a post created in accordance with Galton's will and only changed in name to plain "Professor of Genetics" in 1965.[3]

Galton published this diagram of heredity in *Nature* in 1898, using it to mathematically illustrate how inheritance is passed down through generations.

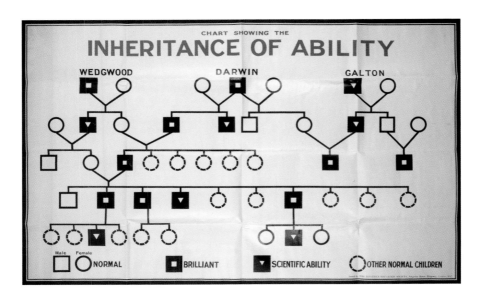

Eugenicists often regarded their own attributes as the most desirable. Ironically, Galton and his wife never had children, so the great man's talents were never passed on to a second generation.

The "Science" of Intelligence

For all its efforts, and the support it had both in intellectual and scientific circles, Galton's movement never achieved legislative power in Britain. The story was different in the U.S. There, industrialists like Andrew Carnegie and John D. Rockefeller decided that eugenics would enable humanity to command its own evolution in a way that was efficient and progressive. In 1904, the Carnegie Institution founded a center for genetic research at Cold Spring Harbor, New York,[4] with Charles Davenport as director. Davenport soon turned his attention to human inheritance. Along with such purely genetic traits as albinism and Huntington's disease, he also traced conditions like alcoholism and "feeble-mindedness" through family lineages, and pronounced these to be Mendelian in nature. In 1910, with the financial support of the Harriman and Rockefeller families, he established the Eugenics Record Office at Cold Spring Harbor and appointed Harry Laughlin as its superintendent.

Meanwhile, the psychologist Henry Goddard had introduced the Binet intelligence test to the U.S. at the beginning of the century. This gave the eugenicists a way to quantify intelligence, and, more particularly, to measure and define what they called "morons," "imbeciles," and "idiots" (in descending order of intelligence). Goddard's famous study of the inheritance of feeblemindedness in the pseudonymous "Kallikak" family was published in 1912. This and other studies showed that the devil of genetic unfitness was at work within the

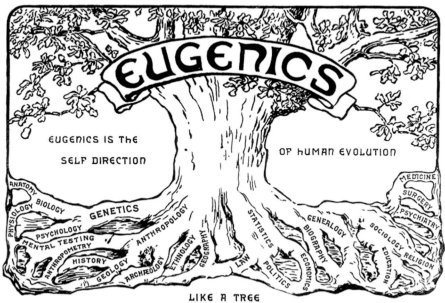

From the early 1900s well into the 1930s the American brand of eugenics inspired state laws mandating sterilization of "defective persons" and federal laws restricting the immigration of the "socially inadequate."

EUGENICS IS THE SELF DIRECTION OF HUMAN EVOLUTION

LIKE A TREE
EUGENICS DRAWS ITS MATERIALS FROM MANY SOURCES AND ORGANIZES THEM INTO AN HARMONIOUS ENTITY.

Gregor Mendel: Planting the Seeds of Genetics

U.S. The only way to deal with the problem was "negative eugenics"—denying the right to breed. By 1931 twenty-seven American states had enacted sterilization laws to allow the compulsory sterilization of certain categories of people, such as the "feeble-minded" and "morons," and by 1941 almost *thirty-six thousand* individuals in the U.S. had been compulsorily sterilized under such laws. The trend spread; within a few years a number of European countries had followed suit with compulsory sterilization, including Nazi Germany, but also Switzerland and the Scandinavian countries.

This retouched photograph of members of the "feeble-minded" Kallikak family served as propaganda during the Eugenics movement.

Race and Immigration

In 1906 the Race Betterment Foundation was set up in Michigan by J. H. Kellogg to promote racial improvement. About the same time, Charles Davenport and Harry Laughlin at Cold Spring Harbor were turning their attention in the same direction. The United States was at that time receiving huge numbers of immigrants of various European "races." Was fine American stock in danger of being contaminated by inferior "germ plasm"? In 1913, in pursuit of this chimera, the psychologist Henry Goddard applied the Binet intelligence test to immigrants at Ellis Island for the first time. On that occasion 80 percent of those tested scored so low as to be considered "feebleminded" (this absurd figure was later revised down, but not by much). Laughlin later appeared as an expert witness before the House Committee on Immigration and Naturalization, and recommended that quotas be introduced restricting the numbers of immigrants from particular, undesirable racial groups. Stringent entry requirements were applied to the fortunate few. Jews were perceived as being just as unfit as any group, and thus many Jews, fleeing racial persecution in Europe, were denied entry to the U.S. by essentially racist regulations.

Anti-miscegenation laws were also enacted. By 1915 twenty-eight states had invalidated marriages between "Negroes and white persons." The 1924 Virginia Racial Integrity Act owed much to the advice of Goddard and Laughlin; it was finally overturned and struck off the books by order of the U.S. Federal Supreme Court only in 1967.

A Transformation

During those years when the entry into the United States at Ellis Island was governed by considerations of eugenics, another, markedly more trivial event took place elsewhere in New York City. Almost every morning a small, elegant man left his apartment on New York's East 67th Street and walked a few blocks to the laboratories of the Rockefeller Institute on York Avenue. Always immaculately dressed in dark jacket and stiff collar, he was almost elfin in appearance, his tiny frame in strange contrast to his large head with its high, domed forehead. Calm, thoughtful eyes looked out on the world through owlish spectacles. He might have been a family lawyer, or perhaps a banker. Certainly none of the people who saw him walk by would have guessed that he was one of the great scientists of the twentieth century, a man who, like Mendel, would usher in a startling new world of discovery. This man was Oswald Avery. If Mendel was the man who first plotted the movement of genes through the generations, Oswald Avery was the man who was to reveal the identity of the gene itself.

There is much about Avery that reminds one of Gregor Mendel. Like Mendel he was a retiring, self-sufficient, confirmed bachelor with a tenacious grip on his research project. Like Mendel, he was a stickler for the exactness of the written word, with a determination not to step beyond the bounds of direct, experimental evidence. Like Mendel, he was one of the greatest experimental biologists in the history of the subject. And also like Mendel—although not to the same extreme degree—in his lifetime he did not receive the full recognition that he deserved.

Known throughout his adult life as "the Professor," or more familiarly, as "Fess," Oswald Avery was born in Canada in 1877, the son of a Baptist minister.

When he was only ten the family transferred from Nova Scotia to the squalor of the most densely populated place on earth, the Lower East Side of New York City, where the immigrants who got through the controls at Ellis Island often ended up. Avery's father was to run a Baptist mission there. It must have been a startling change in the young boy's life, and worse was to come. In the course of the single year of 1892 both his brother Ernest and his father died, probably of tuberculosis, leaving his mother and her two surviving children to fend for themselves. Although the New York Baptist community rallied round to help, in particular John D. Rockefeller himself, life cannot have been easy for the widow and her two surviving children. What effect all this had on young Oswald we don't know, any more than we know directly how Mendel felt about the privations of his own family, for Avery never, ever spoke about his childhood and youth. A gentle, sensitive, and approachable man, much-loved by friends and colleagues, he was nevertheless a deeply private person. No one was ever allowed a glimpse of the inner world that defined him.

The year following his father and brother's deaths, Avery left high school for Colgate Academy. It is probable that he intended to follow in his father's footsteps, for in those days Colgate was a Baptist college training young men for the ministry. However, while he was at university, his life changed direction away from the humanities that he had been studying. Despite having no science education at all, on graduation from Colgate he was admitted to the College of Physicians and Surgeons of Columbia University to train as a doctor. He qualified in 1904 and spent a few years in hospital work but soon moved into medical research. In 1913 he moved to the Rockefeller Institute in Manhattan, where the focus of his work became a single species of bacterium, *Diplococcus pneumoniae*,[5] the agent that causes pneumonia.

For decades, geneticists knew that genes turned hereditary traits on or off, but did not really know how they worked. In 1944 Oswald Avery and his colleagues showed DNA to be the substance of inheritance.

In those pre-antibiotic days the great battles of medicine were against the agents of infectious disease, the bacteria that cause illnesses such as typhoid and tuberculosis. Pneumonia was a major killer, claiming over 50,000 victims a year in the U.S. alone and Avery's pursuit of the bacterium was Mendelian in its single-mindedness: he worked on pneumococcus for almost his entire career of thirty-five years, from 1913 until 1948. For the first decade of that period he attempted to identify the various polysaccharide coats that the organism made, in order to create an active serum to combat the different types of what he nicknamed, with typically gentle irony, the "sugar-coated microbe." Although he achieved some success and a considerable reputation in this work—he was even proposed for a Nobel Prize—he achieved lasting fame for a very different reason.

In 1928 an English microbiologist called Fred Griffith had shown that one strain of pneumococcus, strain **R**, could be transformed into another, strain **S**—and this change was inheritable. **S** was the virulent form that gave the host pneumonia. **R** was a non-virulent form. By injecting mice with non-virulent **R** bacteria and then with heat-treated and therefore dead bacteria of the virulent strain **S**, Griffith had discovered that he could actually induce pneumonia in the mouse *and subsequently recover live virulent bacteria of strain* **S** from the dead animal. This seemed remarkable. One strain of bacterium had been turned into another by something present in dead bacteria[6] and, once transformed, the new characteristic was inherited perfectly normally.

Griffith's paper was not widely read by geneticists or biochemists, but of course it was of direct interest to Oswald Avery in his study of the very same bacterium. To Avery these transformations seemed a disturbing prospect, a problem of pathology rather than genetics. What would it mean for any possibility of fighting the disease organism if pneumococcus could change its nature in this way? So at first—one detects some wishful thinking here—he was inclined to reject Griffith's findings and dismiss them as the result of a poorly controlled experiment. But when members of his group at the Rockefeller Institute repeated the work he couldn't ignore it any longer, and when the same colleagues went on to show that similar transformations could take place *in vitro*—that is, "in glass," outside a living organism—using cell-free extracts from the bacteria, things seemed more interesting still. The implications for pathology slipped into the background behind a more fundamental question: *how could one stable, inheritable strain of bacteria be transformed into another?* Quietly—he was a quiet and thoughtful man—Avery shifted the focus of his research to this curious phenomenon.

Genetic Aberrations—Germany and Russia

As Avery considered transformations in pneumococcus bacteria, in Germany, where genetics research had once been so fruitful, the application of genetic principles to human populations was leading toward a nightmare. As early as 1905, the Society for Racial Hygiene had been founded by Alfred Ploetz. It would be fair to say that the Society was not overtly racist at the outset—indeed Ploetz himself applauded the "Jewish race" as being equal in merit to the "Nordic." Nevertheless, the society's influence was devastating, for when the Nazis came to power in Germany in 1933 they found many of their ideas about human fit-

Gregor Mendel: Planting the Seeds of Genetics

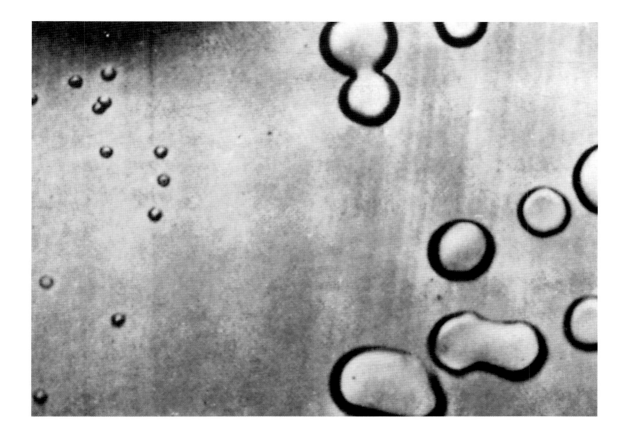

ness already in place within the medical and scientific communities. A eugenic sterilization law began being enforced immediately when Hitler became chancellor and a whole bureaucracy was immediately established—with *Erbklinik* (genetic clinics), *Erbgesundheitsgerichte* (genetic courts), and *Erbämter* (genetic officials). By the 1940s some four hundred thousand people had been sterilized on eugenics grounds—mainly because they were mentally retarded. But in 1939, when the Second World War broke out, sterilization of the mentally retarded was replaced by euthanasia. Now patients in mental hospitals could simply be *killed* on eugenics grounds. Victims of this program, both adults and children, were given lethal injections or gassed; in the occupied territories, they were shot by the same *Einsatzgruppen* that were killing Jews and Gypsies. By 1941, when protests against the policy of gassing patients had become so great that orders were given by Hitler to stop, some 70,000 had been killed. However, killing by other means continued, and even took place in mental institutions after the end of the war.

Parallel with this eugenics program were the Nazi racial doctrines, carried to their hideous conclusion in the concentration camps. It is difficult not to see

Pneumococcus colonies. Microbiologist Fred Griffith studied the inheritability of mutant strains of pneumococcus, the bacterium that produces pneumonia.

As an agriculturalist and plant-breeder, Nikolai Ivanovich Vavilov (left) was an enthusiastic follower of Mendel's principles. Pictured here with Trofim Denisovich Lysenko (right), Vavilov was a star of Russian genetics until his science ran afoul of Soviet politics.

Overview of the Admissibility of Marriage between Aryans and non-Aryans, from the Health Office of the German Reich, 1936. This chart depicts who might marry whom, according to Germany's racial legislation. The white circles represent "pure Germans"; the circles with black indicate the proportion of "Jewish blood."

parallels between Galton's "diagram of heredity" (see p. 121) and the chart of marriages allowed and forbidden under Nazi racial laws. Such racial theories, growing out of the ideas of the Society for Racial Hygiene and other groups, had become an integral part of German academic life in the years following the First World War. The Nuremburg Race Laws of 1935 were drawn up after extensive discussion among leading geneticists.

In Germany Gregor Mendel's enlightenment had become a nightmare; in Soviet Russia things took a remarkably different turn. Rather than being called upon to support racist policies, Mendelian genetics were actually banned.

Nikolai Ivanovich Vavilov was one of the stars of Russian genetics. An agriculturalist and plant-breeder, he was an enthusiastic follower of Gregor Mendel's ideas. In 1913 and 1914 he worked with Bateson, whom he greatly admired, and on a number of occasions he met Thomas Hunt Morgan and other leading geneticists in the United States. Seeing the vital importance of crop plants and plant-breeding to human survival, Vavilov carried out plant collecting missions throughout the world (his most famous book is entitled *Five Continents*) identifying the eight "centers of origin" of cultivated plants. A man of amazing knowledge and remarkable energy, in 1920, at age thirty-three, he was appointed director of the Department of Applied Botany and Plant-Breeding in Petrograd (shortly to be renamed Leningrad, and now, once again, called St. Petersburg), where he set up crop-breeding programs based on Mendelian principles, and established one of the first seed banks in the world. In 1930 he founded the Genetics Institute of the Soviet Academy of Sciences. He was a member of societies throughout the world—for example the American Botanical Society and the Royal Society in Britain—and recipient of a Lenin Prize. He was, in short, the personification of Russian crop-breeding and genetics. And then, during the 1930s, it all went wrong.

As with the eugenics movement in Europe and the U.S., in the Soviet Union there was a fatal cocktail of politics and science. Trofim Denisovich Lysenko was the principal culprit, a man of no scientific understanding who nevertheless possessed a deadly combination of ignorance and cunning. At first he was encouraged by Vavilov but it became clear that what Lysenko was preaching was a kind of Lamarckism, the idea that plant stocks did not improve as a result of acquiring new inherited characters by mutation and artificial selection, but as a result of environmental pressure. Despite being scientifically spurious, such ideas appealed to Soviet politicians, for it was part of the Marxist creed that human beings are fashioned by the environment rather than by inborn talents.

Gregor Mendel: Planting the Seeds of Genetics

Übersichtstafel, betreffend die Zulässigkeit der Eheschließungen zwischen Ariern und Nichtariern.

R.Bu.G. = Reichsbürgergesetz.
G.Sch.d.Bl. = Gesetz zum Schutze des deutschen Blutes und der deutschen Ehre.
AVO. = Ausführungsverordnung.

New Currents in Genetics   129

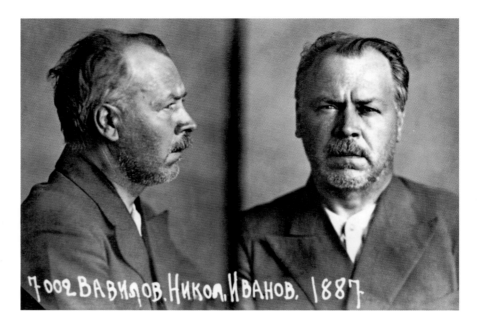

Vavilov was arrested and imprisoned in 1940 for his "controversial" views on inheritance. The political climate of Russia in that era favored Lysenko's views that environmental factors were more influential than inborn talents.

Eventually, labeling Mendelism-Morganism as reactionary and erroneous, "fake products of the Catholic Church and Capitalism," Lysenko gained the backing of Joseph Stalin. Vavilov and others were denounced, Mendelian genetics was banned from the universities and the great plant-breeding establishments were turned over to pseudo-science. In 1940, while on a collecting expedition in Ukraine, Vavilov was arrested. The story has the feeling of Greek tragedy: the throne usurped by a man who has lied and thereby deposed the rightful king. With Vavilov incarcerated, Lysenko became head of the All-Soviet Genetics Institute.

By this time all scientific debate had become victim of a full political purge. Along with Vavilov, dozens of geneticists were arrested and sent to the Gulag. A number were shot. Vavilov himself was sentenced to death in 1941, but the execution was never carried out. By then the full fury of the Nazi invasion had been unleashed on the U.S.S.R. and the appalling two-year siege of Leningrad was underway. During the siege Vavilov's institute covered itself in glory: under the blockade ten members of the staff actually died of starvation while tending the very seed banks that would have provided the nutrition they needed. Meanwhile in 1943 Vavilov himself, the man who had inspired them, died in prison.

Lysenko died of old age as recently as 1967. He is remembered only as the villain responsible for a piece of Stalinist persecution. Nikolai Ivanovich Vavilov, the great Russian Mendelian geneticist, is remembered and revered. The Department of Applied Botany and Plant-Breeding in St. Petersburg is now the Vavilov Institute, still with one of the largest seed banks in the world.

Gregor Mendel: Planting the Seeds of Genetics

Avery and the Discovery of the Gene

While genetics research was mutating into horror in Nazi Germany and being stifled in Soviet Russia, Oswald Avery and his colleagues at the Rockefeller Institute in New York were moving forward with their study of pneumococcus bacteria. From 1934 to 1937 the technique of transformation was perfected, largely through the efforts of Avery's colleague, Colin MacLeod. In 1941, in conjunction with a new arrival at the lab, Maclyn McCarty (MacLeod moved to New York University in 1940) the aim was to isolate and purify the "transforming principle" from the bacterial cells. Systematically the two men set about the series of carefully controlled and endlessly repeated experiments to extract this substance, whatever it was, from one type of bacterium and to purify it. "Some job—full of heartaches and heartbreaks," Avery described it. They treated the extract with many different proteases—protein-breaking enzymes—and found that it was still able to transform one strain into another. This meant that it wasn't a protein. Treatment with lipid-dissolving alcohol and ether did not destroy the transforming ability; so it couldn't be a fat. Therefore they concluded that the transforming substance must be a nucleic acid; yet ribonuclease—an enzyme that digests RNA—did not remove the transforming ability either. The only possibility that remained was that DNA—Levene's dull, uninteresting molecule—was bringing about the transformation. When it was finally demonstrated that deoxyribonuclease enzymes (enzymes that break up DNA) *did* destroy the "transforming principle" the story became clear: the transforming principle must be DNA. "Sounds like a virus—may be a gene," Avery wrote in a letter, in May 1943, to his brother Roy:

> If we are right, and of course that's not yet proven, then it means that nucleic acids are *not* merely structurally important but functionally active substances in determining the biochemical activities and specific characteristics of cells—and that by means of a known chemical substance it is possible to induce *predictable* and *hereditary*[7] changes in cells. This is something that has long been the dream of geneticists ... the problem bristles with implications. . . . It touches genetics, enzyme chemistry, cell metabolism, and carbohydrate synthesis, etc.

Avery signed off: "So there's the story, Roy—right or wrong it's been good fun and lots of work." Good fun it may have been, and lots of work. But it was more than a story—it was epoch-making.

When the results were written up in a joint paper under the authorship of the three men who had been involved in the work since 1934 there was some discussion between Avery and McCarty about the order in which the authors' names should appear on the paper (these details matter to scientists). "I'm not sure," Avery mused, "whether the names go in order of length of association with the research, or in order of age and seniority, or simply alphabetically." McCarty agreed it was a question that had to be resolved. Only when the conversation was over did it dawn on him that he had been victim of one of Fess's quiet jokes: all three options produced precisely the same result: Avery, MacLeod, McCarty.

The paper, published in January 1944 in the *Journal of Experimental Medicine*,[8] is as classic as Mendel's *Versuche*, and equally framed in cautious understatement. After detailing all the experimental evidence and addressing all the possible counterarguments, Avery and McCarty came to their deceptively modest, and circumscribed, conclusion:

> The evidence presented supports the belief that a nucleic acid of the desoxyribose type is the fundamental unit of the transforming principle of Pneumococcus type III.

Thus was DNA first shown to be the substance of inheritance, the stuff of the genes.

It is important to see how revolutionary this research was. Many biologists had despaired of ever being able to isolate genetic material, imagining that a gene would only work as some complex association of molecules, probably protein-based, and that taking such a complex apart would irrevocably and inevitably destroy both the gene's function and any hope of understanding how it did its job. But Avery and his colleagues showed that this was not so. "Terrific and unlimited in its implications," Joshua Lederberg, a future Nobel Prize winner, wrote in his diary when he read the paper.[9]

Yet in other respects the reception of this work was curiously muted. Avery was extremely retiring and unwilling to give public addresses, but McCarty persuaded him to give a lecture on the research to the gathered members of the Rockefeller Institute. "Fess" was met with a reception that echoes that given to Mendel by the members of the Brünn Society for Natural Science: rousing applause, followed by silence. The chairman called for questions and discussion, but there was nothing. Finally one of the senior figures of the Institute rose to his feet and told everyone how long and arduous the research had been and how he admired Avery for his tenacity. Nothing more. The chairman later recalled: "There followed one of those long silences that haunts me yet. Instinctively I felt we were witnesses to something important, even though I cannot say

Gregor Mendel: Planting the Seeds of Genetics

I fully appreciated just how important that paper was to become as the years unrolled."[10]

One reason for the pause before biologists grasped the true significance of Avery's work was that the world was still in the midst of a world war. Peaceful scientific research was restricted in the U.S. and at a halt right across Europe. Another reason may have been the reluctance of some biologists to abandon the idea of protein being the genetic material (there was fierce opposition to Avery's work, conducted by a fellow member of the Rockefeller, Alfred Mirsky), but surely another key factor is an echo of Gregor Mendel's failure to achieve recognition of his own discoveries almost a century earlier. Avery's team published their research in a journal that was read by the wrong people—microbiologists rather than geneticists—and the language of the paper was typically reserved and rather obscure. For example, in the summary (which is what a browsing scientist would read first) there is no mention of the words "gene" or "mutation," or anything much that would have excited the attention of a mainstream geneticist although, in the body of the discussion, they did make this assertion: "The inducing substance has been likened to a gene, and the capsular antigen which is produced in response to it has been regarded as a gene product."

A few significant researchers recognized the importance of this work (apart from Lederberg, see Chargaff, below, p. 136), but contemporary accounts give the strong impression that Avery's discovery was underplayed at the time of its publication and by the time the whole scientific community had come to realize the significance of the "transforming principle" it was too late. In 1948 Avery retired, to spend the remainder of his years living near his brother and family in Nashville, and in 1955 he died. Thirty-five years of research centerd around a single species of bacterium had delivered up "the most seminal discovery of twentieth century biology,"[11] but the author of this work never received the ultimate accolade of the Nobel Prize.

Toward the Double Helix

So the wet blanket of Levene's tetranucleotide theory was thrown aside, even if few people really appreciated it at the time—if DNA could achieve transformations one bacterium to another, then the molecule must be more complex than had been thought. But how complex? What was its composition and what was its shape?

One of the key figures in the next step in the chain of discovery was a British physicist who had spent the last two years of the war in California working on the project to create the atomic bomb. His name was Maurice Wilkins. A

TOP: Maurice Wilkins at King's College took an experimental approach to determining the structure of DNA using X-ray crystallography.

BOTTOM: William Astbury also used X-ray crystallography to study the structure of biological molecules.

tortured young man with a broken marriage behind him and little certainty about his future, he returned to England shortly after the detonation of the bombs over Hiroshima and Nagasaki with one thing clear in his mind: he was determined to move his line of research away from the horrors of war toward the mechanisms of life. Partly inspired by the book written by the quantum physicist Erwin Shrödinger entitled *What is Life?*, Wilkins felt that he could use his knowledge of physics to explore those molecules that were on the borderline between the living and the non-living—the molecules of genes and chromosomes. At the biophysics laboratory at King's College London, he first attempted to use ultraviolet light to try to follow the movement and proliferation of DNA in dividing cells (DNA absorbs UV light), but his attention soon shifted. X-ray crystallography[12] was a means of exploring the actual structure of a molecule. This technique had been used with DNA just before the war by William Astbury at the University of Leeds in England, but Wilkins decided to attempt to perfect it. During a scientific congress in London he obtained some excellent DNA samples from a German biochemist—it was a kind of goo that looked, so one of the scientists proclaimed, "like snot"—and he set to work on this. By 1950, in conjunction with a Ph.D. student, Raymond Gosling, and with the help of a physicist at the laboratory, Alec Stokes, Wilkins had come up with the idea that the molecule was helical in shape. This was a start, but he realized that they needed more expertise in X-ray diffraction techniques and he pressed the head of the laboratory to find one. In January 1951 the expert arrived. She was Rosalind Franklin, a powerfully intelligent young woman who had spent the previous four years working in Paris. Forceful and fascinating perhaps sums her up. Dressed in the Dior "New Look" style that was all the rage, she was certainly not the aggressive frump that James Watson describes in his book *The Double Helix*.

Returning to her native London after the years in Paris, Franklin was afraid that she had made a dreadful mistake in returning. The city was still in the midst of post-war reconstruction, rationing was still in place, and what she found at the King's College did nothing to reassure her. The place was partly a bomb site (they were building a new laboratory in one bomb crater) and was

Gregor Mendel: Planting the Seeds of Genetics

old-fashioned in its manners (there was, for example, a "men only" common room). For his part Wilkins, a self-effacing man who did not relish argument, considered her a daunting colleague who appeared reluctant to share her ideas. But there was no doubt about the quality of her experimental ability. Settling down in a cramped basement lab she began to work at setting the thin fibres of DNA up for irradiation. The problem with using X-ray crystallography on DNA stems from the fact that, as the name implies, this technique was invented for analyzing the structure of crystals. Crystals are built up of repeating, regularly arranged layers of atoms and it is this regularity that diffracts the X-rays and gives the diffraction pattern that enables a physicist to calculate the positions of atoms within the crystal. DNA simply isn't crystalline. It doesn't naturally have regular layers of atoms, so the trick was to draw the "goo" out into fine threads rather like the threads of mozzarella cheese you get when eating pizza. This would, it was hoped, achieve a sort of approximate crystalline structure with all the molecules lined up parallel to one another in the thread.

Rosalind Franklin's X-ray diffraction photographs paved the way for James Watson and Francis Crick's final determination that the structure of DNA was a double helix, like a twisted ladder.

Painstakingly, Franklin experimented with different levels of hydration of the threads. Long exposure to X-rays—up to one hundred hours—was needed in order to achieve a decent pattern, and once this had been done she had to interpret it. The pictures themselves are strange patterns that might have come from the brush of an abstract painter, shimmering geometries whose hidden meaning may only be teased out of them by complex mathematics and a subtle intuition that seems to be more the province of the artist than the scientist. Nowadays computer programs can cut the calculation time down to seconds; then it took hours and hours with a slide rule and notebook.

Sadly, Franklin did not communicate her progress to the very man who had originally called for the employment of an X-ray expert, Maurice Wilkins. Although initially polite, their relationship steadily deteriorated. On one occasion, after he had given a particularly successful lecture on their DNA findings to a meeting of scientists in Cambridge, she even came up to him and demanded that he cease his work on DNA. "Go back to your microscopes!" she said sharply. Wilkins was understandably bewildered by this response. A major part of the difficulty appears to lie in the fact that Wilkins thought of Franklin as his assistant, but in reality and quite unknown to him, the head of the laboratory had given her the clear impression that she would be working on her own. Thus she saw interference by an outsider where Wilkins intended interest and collaboration.

Meanwhile, inspired by Avery's paper on bacterial transformation, Erwin Chargaff at Columbia University had set to work on a careful analysis of nucleic acid. He was able to use the newly developed techniques of paper chromatography and ultra-violet spectrophotometry to obtain a much more precise assay of nucleic acids than had been possible before from a variety of sources ranging from calf thymus to yeast. In 1950 he published an accurate analysis of the composition of DNA that once and for all put an end to Levene's tetranucleotide theory of repeating units. The bases adenine, guanine, cytosine, and thymine were *not* present in equal numbers. Although there was always as much A as T, and as much G as C (suggesting some kind of pairing) he found that the A + T/G + C ratio was constant for any one species that he tested, but actually differed from species to species. He grasped the implications of this:

Desoxypentose nucleic acids from different species differ in their chemical composition . . . and I think there will be no objection to the statement that . . . they could very well serve as one of the agents, or possibly as the agent, concerned with the transmission of inherited properties.[13]

This 1952 X-ray crystallography image of DNA was the key to the discovery of DNA's structure. Franklin got this image by directing X-ray beams at DNA molecules, whose atoms diffract the beam differently.

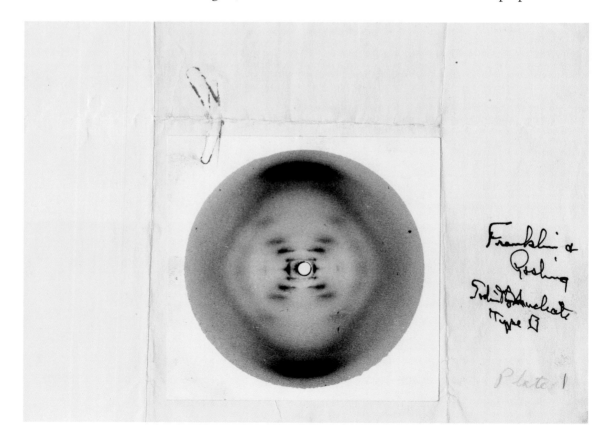

Gregor Mendel: Planting the Seeds of Genetics

The hunt for the structure of DNA was truly underway. As further confirmation of its role, two workers at Cold Spring Harbor were in the process of showing that when a virus infects a bacterium it is only the virus's DNA that causes the infection. The viral coat—entirely protein—remains outside the bacterial cell and plays no part in infection. So DNA was clearly the molecule that mattered. Linus Pauling, the renowned protein chemist in California, joined the race against Franklin and Wilkins in London, and so too did a rather maverick couple working at Lawrence Bragg's laboratory at Cambridge University. This unlikely pair—James Watson, a brash young American biologist who had started his scientific career as an ornithologist, and Francis Crick, an outspoken British physicist—knew of the work at King's College in London; indeed Crick was an old friend of Wilkins's and was present at the latter's talk on DNA mentioned above. Both Crick and Watson and the King's group were funded by the same organization, the Medical Research Council (MRC), so it was natural that there should be an exchange of views between them. In contrast to the way Wilkins and Franklin were working, the approach of Crick and Watson was highly speculative. Being a physicist, Crick was able to interpret X-ray diffraction patterns, but they themselves did no experimental work at all. What they did do was follow Pauling's lead and attempt to build molecular models. Their first attempt—Franklin and Wilkins were called up to Cambridge to see it—was a complete flop,[14] but in early 1953 they were handed the means of solving the problem. By this time relations between Wilkins and Franklin had become so strained that Rosalind Franklin was preparing to leave King's for another post in London and when she was putting things in order for her departure, Franklin's Ph.D. student Raymond Gosling handed Wilkins an X-ray diffraction image. Apparently the photo had been sitting in a drawer in Franklin's desk for months.[15] It was, quite simply, the best image yet. Even a non-expert could see that there was more information there than in any of the other pictures that had been produced: more detail, more focus, more precision. Gosling seems less certain than Wilkins about this photograph. In an interview on a PBS program entitled "The Secret of Photo 51" and aired on April 22, 2003 he says: "I cannot remember how he [Wilkins] came by this beautiful picture. It may have been given to him by Rosalind, or it may have been me." However Wilkins obtained it, there is no doubt what happened next: Wilkins showed it to Watson. It is difficult to see why he should not have done so, but he later admitted regretting the move, and certainly from Watson's

Erwin Chargaff, a researcher at Columbia University, in 1950 published an accurate analysis of the composition of DNA.

James Watson (left) and Francis Crick (right) announced their double helix discovery in the April 25, 1953 issue of *Nature*. Separate papers by Franklin and Wilkins, based on their X-ray diffraction work on the molecular structure of nucleic acids, appeared in the same issue.

own account this photograph was vital in spurring the young American to return to Cambridge and get back to model building. He might not have understood X-ray crystallography but he knew the signature of a helix when he saw it. There was also another source of information that was directly connected with Franklin's work, a report to the MRC that included Franklin's calculations of the dimensions of the molecule. This was made available to Crick by Max Perutz, who was the head of the unit where they worked.

Back in Cambridge, Watson and Crick pondered this information, scratched their heads over Chargaff's work and a few other clues—Jerry Donohue's information about the chemical nature of the bases was crucial—and constructed their DNA model. The result—startling, logical, and possessing a cool sculptural beauty—was the now-iconic double helix. On that morning of February 28, 1953 the solution to the problem must have seemed an epiphany. It was hardly surprising that at lunchtime the pair went into the local pub, the Eagle, and the voluble and excited Francis Crick made his famous announcement to the startled customers: "We've just found the secret of life!"

"It has not escaped our notice," they later wrote, more soberly, in their paper in *Nature* of April 1953, "that the specific pairing [of bases across the axis of the molecule] we have postulated immediately suggests a possible copying mechanism for the genetic material."[16]

Gregor Mendel: Planting the Seeds of Genetics

Mendel's *Elemente*, once nothing more than an idea, had finally acquired a firm chemical identity. Wilkins was offered a share of the authorship of the paper in *Nature*, but declined. His own paper and Franklin's (both highly technical) were published separately in the same edition of the journal and he continued to work on the molecule for the next few years, confirming and adjusting details of the original model. "I loved DNA," he asserts in his autobiography. "I wanted to savor its nature and find what that nature revealed." Perhaps it is the same with all biologists, indeed, all scientists. Maybe Mendel loved his peas, and Morgan his fruit flies. It is that kind of love that may kill the beloved, but it is love nevertheless. Maybe Avery loved the pneumococcus bacterium to which he dedicated his life—certainly he would never be present in the lab when his colleagues were performing the heat destruction of liters of cultured bacteria that was necessary for obtaining the "transforming principle."

When finally the discovery came up for the Nobel Prize in 1962, Watson, Crick, and Wilkins were credited equally with the discovery. It will never be known whether Rosalind Franklin would have been honored alongside them; she died of ovarian cancer in 1958 at the age of thirty-seven, the disease probably brought on by excessive exposure to those very X-rays that had revealed the structure of the molecule.

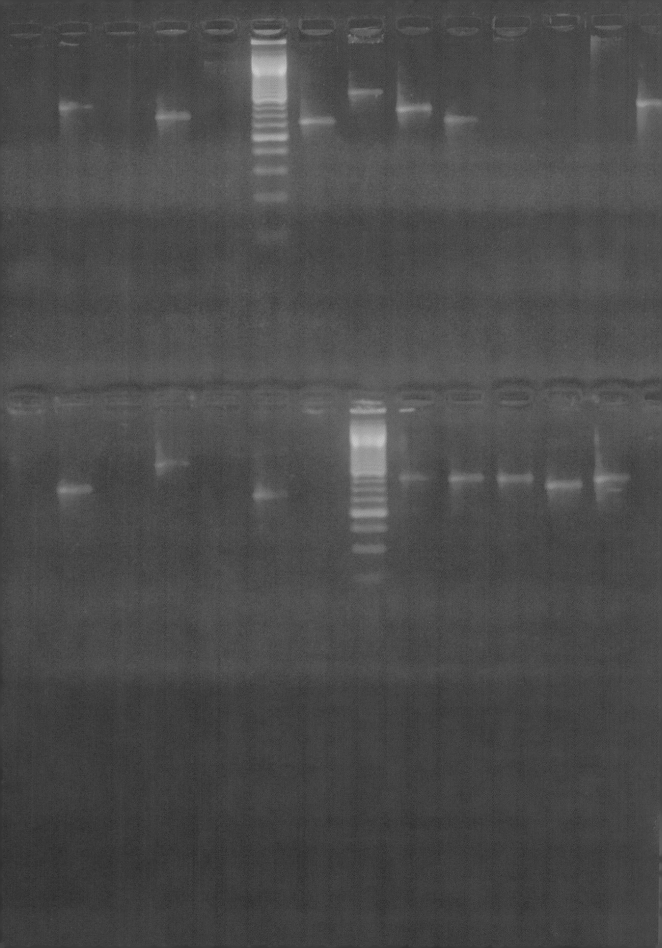

From Code to Genome

Once James Watson and Francis Crick had discovered the structure of DNA the doors to discovery were flung open. By now it was known that protein synthesis, the assembling of small amino acids into the long chains of the finished protein molecule, takes place in the cell's cytoplasm. However DNA, Avery's "transforming principle," resides in the nucleus of all cells other than bacteria,[1] so the next question was, how do the two link together? How does DNA, which somehow holds the genetic information, get its message out into the cytoplasm where minute cell inclusions called ribosomes actually churn out the protein chains? In particular, what was the role in all this of the other nucleic acid, the one that Avery had shown was *not* the "transforming principle," the one called RNA?

A brilliant and versatile Russian physicist called George Gamow became one of the mentors of this discussion. Gamow was big in every way—huge physically and powerful of personality, and with an impressively explosive reputation: he was the man who first proposed the Big Bang theory of the origin of the universe. A marvellous popularizer of science (his Mr. Tomkins books are among the most entertaining ways of understanding such rarefied matters as relativity and quantum mechanics), he now turned his attention to the problem of what he called the "genetic code," the correspondence that there must be between the DNA bases and the amino acids of protein. In 1954 he founded the small, elite discussion group known as the RNA Tie Club. Numbering a mere twenty members (one for each of the twenty amino acids in proteins) and four honorary members (one for each nucleotide in nucleic acid) this informal group corresponded over the latest ideas and discoveries concerning DNA and RNA and protein synthesis.

In a common sequencing process, DNA is immersed in a gel and charged with an electric current, which causes the DNA fragments to separate by size. The gel is then stained with a chemical that binds the DNA and glows when exposed to ultraviolet light, allowing us to see the DNA fragments more clearly.

The members were entitled to wear a necktie of horrendous aspect designed by Gamow himself—it consisted of an embroidered helix down its length along with representations of the nucleic acid bases—and each member had an amino acid ascribed to him. Thus, Francis Crick was "Tyr," for the amino acid tyrosine, whereas James Watson was "Pro," for proline. Out of the twenty members of the club eight were or became Nobel laureates, which gives an idea of the intellectual power of the group.

Throughout the latter half of the fifties and into the sixties, letters went back and forth between members of the club, with Crick often leading the discussion. Indirect evidence suggested that RNA was the messenger between the DNA of the nucleus and the ribosomes of the cytoplasm, but what was the nature of the coded information and how did the ribosomes read it? Another problem was the genetic code itself. It was apparent that the DNA letters (the four nucleotides represented by the letters A, C, G, and T) would have to be read in groups of at least three (so-called triplets) in order to provide enough "words" for the twenty amino acids, but what were the words? What combination of DNA letters meant, for example, Crick's amino acid tyrosine or Watson's proline? In 1955 Crick suggested that the cell would need what he called an "adaptor molecule" to read the message off the RNA during the translation process, a molecule that would follow the base-pairing rules that Chargaff had discovered and he and Watson had used in their building of the DNA molecule: A with T and G with C.[2] But it was a scientist quite unknown to the RNA Tie Club who cracked the problem of the code itself.

Marshall Nirenberg started out as a biologist before he moved into the rarefied world of what was coming to be called molecular biology. Indeed, as a teenage naturalist he drew spiders and insects and recorded the plant life around his home. His bachelor's degree was in zoology and chemistry but while the thesis for his master's degree was on caddis flies, his interests were shifting toward the biochemical. When he moved to the University of Michigan at Ann Arbor in 1952 the transition was complete. His doctorate in 1957 was granted for research into the uptake of hexose (six carbon sugars) by tumor cells; by 1959 he was at the National Institutes of Health in Bethesda, working on the interaction between DNA and RNA.

In collaboration with a post-doctoral researcher from Germany, Heinrich Mattaei, Nirenberg developed a method that was both elegant and ingenious to solve this question. Firstly they obtained an extract of the cytoplasm of the common bacterium *E. coli* that contained the protein-making apparatus of the cells—minute sub-cellular organelles called ribosomes. Then, on the assumption that RNA was indeed the messenger, they simply wrote their own message. By making the RNA with just one base, Uracil (equivalent to the T—thymine—of DNA), they created a message that just said one "word" to the cell extract—the word UUU. They then fed this artificial message to the cell extract and looked to see which amino acids would be incorporated into the resulting protein. Beautifully simple, the technique worked. Given the message UUU the ribosomes incorporated the amino acid phenylalanine into a chain. To the cell the word UUU "meant" the amino acid phenylalanine. The first word of the genetic code had been read.

Gregor Mendel: Planting the Seeds of Genetics

In 1961 Nirenberg and Mattaei published this finding in a paper in the Proceedings of the National Academy of Sciences. The same month Nirenberg traveled to Moscow to present their research at the International Congress of Biochemistry.

Nirenberg was almost entirely unknown to the scientific world when he delivered his lecture and there were no more than thirty biochemists in the audience. However, also at the conference was that intellectual catalyst and RNA Tie Club member, the one whose amino acid was tyrosine, Francis Crick. Crick had heard about Nirenberg's work and grasping its significance, he immediately rearranged the schedule of the entire congress. He himself would chair a special plenary session where the quiet American could repeat his lecture. This time Nirenberg spoke to an audience of nearly a thousand and by the end of his talk, he had been catapulted to scientific fame.

Marshall Nirenberg was catapulted to international fame when he published his findings on the genetic code.

It is ironical that Moscow should have been the scene for this next milestone in the history of genetics. Although Leningrad and Vavilov's institute had survived the Second World War, so too had Lysenko and his ideas. Lysenkoism was only finally being discredited in the late 1950s and Mendelism itself was gradually being reinstated in the universities of the U.S.S.R. and its satellites. Now, with congresses such as the one Nirenberg attended, Russia was trying to put herself back onto the scientific map. When he returned to the United States, Nirenberg received a congratulatory postcard from the Russian biologist Zhores Medvedev[3] thanking him for his work on the cracking of the code. Medvedev wrote, in occasionally uncertain English, "Biologists and geneticists [in Russia] for the very long time criticized and rejected the coding problem as non-scientific and idealistic (Morganism-Mendelism). You probably know about long and great discussion about heredity that was started by Lysenko and which seemed to be permanent. The new experimental direct attack in coding problem is a good basis for the settlement of this hard discussion."[4] The scientific Cold War, if not the real Cold War, was thawing.

With the publication of Nirenberg and Mattaei's work, the deciphering of the genetic code had only just begun. Their technique was out in the public domain now and other workers, in particular the Indian biochemist Gobind Khorana at the University of Wisconsin, joined in the attempt to decipher the whole code—all the other RNA "words" for the twenty amino acids that are incorporated into protein. Over the next five years Nirenberg, now working with a team of researchers, managed to keep up with the pack and by 1966 was able to announce that he had the solution to the whole genetic code—all sixty-four

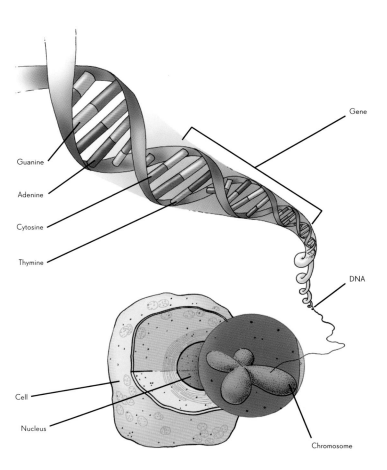

Guanine

Adenine

Cytosine

Thymine

Gene

DNA

Cell

Nucleus

Chromosome

Genes are made up of a "stretch" of DNA that exists inside each chromosome.

codons (the three-base "words") and the amino acids they represented. Exactly a century after the publication of Mendel's paper, the words of the genes could now be read.[5]

"Genetic code" is a wonderful and mysterious phrase, replete with the promise and excitement of science; except that now it had become "language" and mere high school students could learn how DNA works: as you move along the molecule you come across a series of "bases" (those that Kossel identified all those years ago). The bases are given the letters A, G, C, and T from their chemical names. The message that they hold is read in groups of three, every "triplet" representing one of the twenty amino acids (some amino acids are represented more than once). This correspondence of bases to amino acids is Nirenberg's genetic code. Thus the sequence of DNA bases CACGTAGACTGG codes for the amino acids:

histidine—valine—aspartic acid—tryptophan.

Gregor Mendel: Planting the Seeds of Genetics

However, Nirenberg and others were merely reading the words of the genetic language. What about trying to read entire genes? The genes were clearly *there* on the chromosomes—Morgan and his co-workers had shown that; they were made of DNA—Avery had shown that; and now the vocabulary of the molecule was known. But how could a gene actually be isolated? And how could its sequence of words be read? It was as though we had got a list of words for a language but had no idea how they were put together to form sentences.

For a short while it seemed that progress was stalled. What was needed were the right molecular tools.

The Tools of the Trade

A major step was the discovery that bacteria already did their own genetic manipulations, and that we could use their equipment—the so-called "restriction" enzymes. These enzymes (hundreds of different ones are now known) cut DNA into fragments at very specific sites—particular sequences of base pairs. To sort out the fragments there is the technique called electrophoresis, where an electric current passing through a slab of gel is used to separate the different-sized DNA fragments with great precision. Then there are other bacterial enzymes called ligases that give the biologist the ability to stitch the DNA fragments back together in the way desired. It is safe to say that without bacteria no progress at all could have been made with this work, for apart from these enzymes bacteria also carry with them short loops of DNA called plasmids, which may be used as carriers for chosen genes from another organism.

Using these tools you can, for example, clone a human gene: you can cut up human DNA with a particular restriction enzyme, separate the fragments using electrophoresis, and identify which one is your gene, then cut up a bacterial plasmid with exactly the same enzyme and insert your chosen DNA fragment into the plasmid using the ligase enzyme. Finally, you put the plasmid back into the bacteria and grow them in culture. As the bacteria multiply, so also do the plasmids that they are carrying, and there in the plasmid is the human gene that was chosen at the start—in fact you are making multiple copies, clones, of the gene. You may even be able to persuade the bacterium to express the gene that you have inserted—that is, get the gene to work in a completely different cell from the one where it originally operated. This, for example, is how human insulin can now be produced commercially.

While developments like these gave molecular biology the basic tools, within a few years two further breakthroughs were made. One came after years of careful and systematic research and brought with it a second Nobel Prize to the Cambridge biologist Frederick Sanger. The other came in a flash of inspiration.

Sequencing and Multiplying

In 1958 Frederick Sanger at Cambridge University in England had already won a Nobel Prize for working out the arrangement of amino acids in the protein insulin; by the 1970s he had turned his unwavering attention to DNA and before the end of the decade he had developed a way of actually sequencing DNA—reading the bases in order—and had used it to read the first entire genome,[6] that of a virus with the arresting name of phi-X 174. At last whole sentences of DNA could be read. Even better, his method lent itself to automation. The way was open for biologists to move on to read complete books—the genomes of whole organisms.[7]

Sanger's second Nobel Prize, awarded for this sequencing technique came in 1980 and was the fruit of many years' meticulous work. Kary Mullis's prize came in 1993 and was the product of a moment's insight. A chemist working in the new biotechnology industry, Mullis was driving one night in April 1983 to his cabin in northern California. On the way, so the story goes, the idea for the Polymerase Chain Reaction came to him in a sudden epiphany. Ignoring the protests of his girlfriend trying to snatch some sleep in the back seat, he pulled the car over to the side of the road and scribbled the idea down there and then. Mind you, two years later he held a conversation with a glowing, talking racoon outside the same cabin, so he doesn't always get things right. But on that first night he did, for it is hard to exaggerate the importance of PCR. It revolutionized molecular biology. Using a PCR machine (the process is easily automated), in a matter of hours a precise piece of DNA can be amplified a million-fold. You have to know two specific short sequences on either side of the target DNA and you have to make two DNA primers to fit those two sequences (that's where the "gene machine" comes in), but once you've done that the rest is easy. The technique is absolutely precise, quick, straightforward, and is used all the time, every moment of every day, wherever molecular biology is being done. Employing Sanger's sequencing technique laboratories can now read entire genomes; using Mullis's PCR they can make hundreds of thousands of copies of a gene—or any desired section of DNA—in a few minutes. Put the two together and have dozens of centers all over the world working collaboratively and you have the Human Genome Project, which delivered its draft version of the full human genome—the whole length of human DNA—in 2000, exactly one hundred years after the rediscovery of Mendel's paper.

What about the Gene?

This whole story started out with Gregor Mendel, and it has followed the progress of the gene, the *Elemente* that Mendel perceived indistinctly through the

Gregor Mendel: Planting the Seeds of Genetics

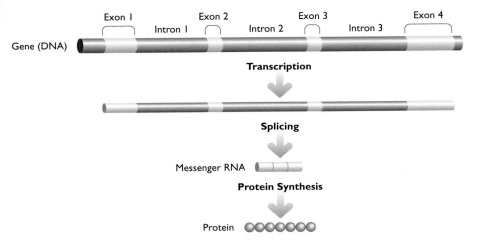

	Exon 1		Exon 2		Exon 3		Exon 4
		Intron 1		Intron 2		Intron 3	

Gene (DNA)

Transcription

Splicing

Messenger RNA

Protein Synthesis

Protein

Scientists currently define a gene as a DNA sequence that codes for a specific trait. This sequence is made up of coding regions (the exons) and noncoding regions (the introns). During transcription (turning DNA into RNA), the introns are spliced out, leaving only exons in a continuous stretch. This "messenger RNA" provides the template for protein synthesis.

pattern of inheritance in the garden pea. So the question now arises: what are genes to modern biologists? They are certainly strung together on the chromosomes (twenty-three of them in humans) but they are no longer thought of as Morgan's "beads on a string"; indeed, they no longer even have the *unity* that Morgan and everyone else imagined right up until the developments in molecular genetics in the 1980s.

We now know that the actual "sense" part of many genes may be broken up by lengths of non-coding DNA known as "introns"; but more than this, genes themselves are separated from one another by vast lengths of DNA that don't appear to do anything at all—so-called junk DNA.[8] And now, after a cooperative effort among researchers throughout the world, the whole human genome, genes as well as non-coding sequences, has been read. The length of it all is difficult to imagine. In all, there are 3.3 *billion* DNA base pairs in the human genome, divided across the twenty-three chromosomes. As only a tiny fraction of these actually make up functioning genes (2 to 3 percent), the remainder of the DNA seems only to give length—and the length is staggering. Superlatives just won't do. In any nucleus of any cell, the twenty-three chromosome pairs contain a total of just under *two meters* of DNA. All of that length, a minutely thin thread, is spooled onto thousands of protein beads, and wound up and stacked until it is manageable—the chromosomes that Boveri and Sutton first saw and drew. There are about fifty trillion (5×10^{13}) cells in the human body. If you unravelled the DNA from all these cells—you can do it easily enough for a few but doing it for all of them might be a bit tedious—you would end up with a total length of DNA about one thousand million, million kilometers long. The furthest planet, Pluto, is a mere six thousand million kilometers from the sun, so all the DNA

Young Barbara McClintock, at Cornell University, demonstrated for the first time that genetic recombination is accompanied by an actual physical "crossing over" between paired chromosomes.

from a single human would *stretch out from the sun right across the solar system to Pluto and way, way beyond.* That isn't a factoid; it's a *fact.*

And yet—and this is possibly even more startling—from this immense length of DNA a molecular biologist can cut out one of those genes using a restriction enzyme, isolate it from all the other fragments using electrophoresis, stitch it into a bacterial plasmid using ligase and grow the bacterium into a few million copies. Thus you can build up gene libraries. They exist in laboratories around the world, in bacteria (the gut bacterium *E. coli* is the species of choice), as well as in viruses and yeast cells. And not only libraries of human genes: plant genes have been cloned in the thousands, as well as genes from numerous different animals. And the complete genome has been read and recorded from organisms from yeast to the domestic dog.

There is a further surprise. If you happen to open any biology textbook published *before* the year 2000 you will find an estimate of around one hundred thousand for the total number of human genes. But when the draft of the full human genome was published, the number of genes was dramatically reduced—apparently we really don't have many at all: perhaps fewer than twenty-five thousand. And yet we probably make some ninety thousand proteins, which means that some genes can ultimately make more than one functioning protein each—Beadle and Tatum's one gene: one enzyme hypothesis (subsequently modified to one gene: one protein polypeptide) has been shaken.

This rather limited number of genes leads to a branch of genetics that promises to occupy much of the new century. It goes under the name of epigenetics—how do those twenty-five thousand genes turn each other on and off, how do they make the patterns in development that become tissues and organs and ultimately the whole organism? How does a chimpanzee become a chimpanzee, a human a human, and a sea squirt a sea squirt? Why are there "conserved" sequences of DNA that are identical in a whole range of distantly related organisms? For example, why do we now discover that Morgan's famous fruit flies have almost exactly the same developmental genes in exactly the same sequence along a particular chromosome as both mouse and man?

There's a further phenomenon that upsets the beads-on-a-string hypothesis. We owe its discovery to a scientist at Cornell University named Barbara McClintock. McClintock's first major contribution to the subject had been in

McClintock, pictured here with Almiro Blumenschein and Kato Yamakake, dedicated much of her genetics work to the study of corn.

the late 1920s when she and a graduate student named Harriet Creighton demonstrated for the first time that genetic recombination (see p. 112) is accompanied by an actual physical "crossing over" between paired chromosomes. Then over the next few years she made the most remarkable discovery, that some bits of DNA (termed transposable elements, or transposons, or even—not a good name—"jumping genes") are capable of moving from one chromosome to another. Such capricious behavior was met with stony disbelief from the scientific world and it was not until the rise of molecular biology in the 1970s that biologists began to discover other such transposable elements were also present in bacteria and yeast, in animals, and in higher plants. McClintock's time had finally come; in her later years she received full recognition, culminating in the award of a Nobel Prize in 1983.

Where Do We Go from Here?

The obvious applications of the new genetics, the ones that grab the front pages of the world's newspapers, are in medicine and forensics. Who has not heard about possible cures for genetic disease, or DNA "fingerprinting"[1] being used to bring a criminal to justice? But molecular genetics has become the driving force in all aspects of biology. Surely this would have given Gregor Mendel—a great field biologist himself—as much pleasure as the more obvious medical advances. In Mendel's day (and starting earlier with the Swedish biologist Carl von Linné[2]) biology was largely comparative. Biologists compared plants and animals anatomically, and put them into groups based on such comparisons. It was rather like stamp collecting—indeed the father of atomic physics, Ernest Rutherford, implicitly criticized biology for that very feature: "Science is either physics or it is stamp collecting," he said in a famous put-down. Perhaps Rutherford was being unfair, for evolution had long ago given biology a framework on which the hundreds of thousands of plant and animal groups could be hung; nevertheless, the great physicist had a point. Based on their own prejudices and opinions, biologists tended to argue animals and plants into groups of their choice rather as though stamp collectors were arguing about how to arrange stamps in an album, whether by shape or value or country of origin. Bizarre theories abounded—for example, that vertebrates evolved from arthropods. They are, after all, both segmented animals and segmentation is a very fundamental feature of body design. However, the arthropods have their main nerve cord along their bellies whereas vertebrates have theirs along their backs—so the neat solution was the suggestion that arthropods must have lain on their backs in order to evolve into early vertebrates! There was no concrete way of disproving this kind of circular argument until the systematic analysis of family trees was developed in the 1960s, a science known as cladistics. In cladistics, features of a group of living organisms are analyzed and a tree (or cladogram) is drawn up that shows the relationships of these organisms based on those features. If the analysis of the features is correct, this cladogram will represent the evolutionary tree by which the organisms are related. Usually the analysis seeks the simplest pathway, with the fewest evolutionary changes, that would

The Tree of Life project is using genetics to document the familial relationships of the 1.8 million known species that exist on earth.

A modern genetics lab is a resource for scientists from all disciplines who study genetic relationships among plants and animals. Each year the Field Museum's Pritzker Lab for Molecular Systematics and Evolution collects over 70,000 sequences from organisms ranging from fungi to plants to birds to insects.

account for the evolution of these features, a process known as "parsimony analysis"—the thought being that the "leanest" family tree is the most likely to be the correct one. But creating family trees using visible features was always difficult, partly because visible features are always subject to natural selection and therefore change from one group to another. Markers that would not be affected by evolution itself were needed, and DNA provides just such markers. When DNA is replicated the occasional mistake is made—a mutation. Such mistakes occur at a fairly constant rate and they accumulate over time, particularly in "neutral" DNA, those parts of the total DNA that don't have any effect on the organism's survival. Such mutations are needed as markers for working out family trees. The more of such markers two organisms have in common, the closer together they must be in the evolutionary tree. So while cladistics introduced objectivity and rigor, the advent of molecular genetics introduced a better precision into the process. By comparing similarities and differences of DNA markers, combined with morphological features, scientists can plot the tree of life and trace the family relationships of all organisms on earth.

The Tree of Life

The Tree of Life is a project to work out the family relationships of all known species on earth—about 1.8 million at the current count (although it is esti-

Gregor Mendel: Planting the Seeds of Genetics

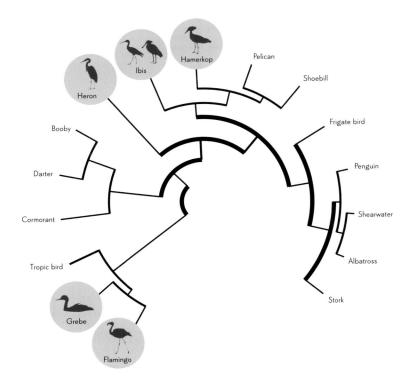

Genetics is transforming taxonomy. The flamingo was once classified with similar long-legged waders such as spoonbills, but DNA analysis now shows that the flamingo's closest relatives are grebes, short-legged diving birds.

mated that there are about thirteen to fourteen million in total). It is a vast collaborative effort among some twenty-five institutions in the U.S. and abroad. As Dr. Shannon Hackett, one of the participants in the Tree of Life project and a researcher at the Field Museum of Chicago says, "The same way you're interested in your own genealogy—your relationships with your parents, your grandparents and your great-grandparents—those are the same kinds of concepts that we're interested in, but across much longer timescales. So, not just how individuals might be related to each other, but how populations, species, classes, or even phyla are related to each other."

This remarkable project is possible today for two reasons. Firstly, the Human Genome Project stimulated advances in automated DNA sequencing so that nowadays, with the very latest techniques, a complete bacterial genome may be read in a matter of days—and the process is getting faster. The human genome, with its three billion base pairs is about a thousand times larger than that of a bacterium, but even so the new techniques will soon be able to read a complete animal or plant genome in a few months. Secondly, the vast amounts of data generated by these sequencing techniques can only be handled because the equally vast amount of computing power is now available to store, analyze the data statistically, and draw up the family trees of all the groups of living organisms.

Within the Tree of Life project, Hackett and her colleagues are focusing on the phylogeny of the birds. The work has already produced surprises. For example, the flamingo, previously classified in the same group as the rather similar long-legged waders such as spoonbills, is actually not close to those birds at all. Instead its closest relative are grebes, short-legged diving birds. The point is that the physical structures—the shape of legs or beaks or some other aspect of anatomy—that biologists have used up to now are only the secondary expression of the underlying genes. Now we can compare the genes directly. In other words we are getting at the true evolutionary relationships that lie beneath the sometimes deceptive morphology of animals and plants.

The bird lineage comprises only a part of the Tree of Life—the whole project includes all the other vertebrate animals, as well as the invertebrates (there are three hundred thousand beetle species alone!); it includes all the plants, from single-celled algae to three-hundred-foot-high giant sequoia trees; all the fungi, from molds and yeasts to mushrooms; and all the bacteria. Altogether it is a truly monumental undertaking, greater is size and scope than the Human Genome Project itself.

Scientific collections are invaluable for mapping genetic variation within and among species. These tanagers are from the collections of The Field Museum, Chicago.

Gregor Mendel: Planting the Seeds of Genetics

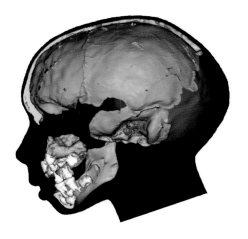
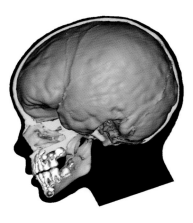

The reconstructed skull on the left is that of a Neanderthal child who lived more than 100,000 years ago. On the right is a skull from a modern human child of comparable age. DNA analysis indicates that the closest common ancestor of Neanderthal man and modern man lived more than 600,000 years ago, proving that they are different species.

Who Are Our Ancestors?

Another key aspect of this is the so-called molecular clock. Using the molecular clock we can work out how long ago certain key events in evolution occurred. This is because mutations appear to occur at a fairly constant rate, and as they accumulate over time in the "neutral" DNA they can be used to estimate the elapse of time. For example, over a period of a million years, there will be roughly a 2 percent divergence in mitochondrial DNA[3] between two related species. So if you have two modern species and you find that a stretch of their mitochondrial DNA differs by 4 percent, then they must have been the same two million years ago—so that is when the two modern species must have had a single common ancestor.

The useful thing about mitochondrial DNA is that it comes to you directly from your mother, because she is the only contributor of the mitochondria to the fertilized egg. The chromosomal DNA of the nucleus undergoes "mixing" each generation, parts of each maternal chromosome crossing over with parts of each paternal chromosome (see *crossing-over* and *recombination*, p. 112), but there is no such mixing of DNA in mitochondria. You get your mother's mito-chondrial DNA complete and unadulterated. Because of these curious features, mitochondrial DNA may be used to trace female lineages. One of the most cel-ebrated examples of this was in 1997, when Brian Sykes, Professor of Human Genetics at Oxford University, showed that the so-called Cheddar Man, a Stone Age skeleton found in 1903 in a cave in Cheddar Gorge in southwest England, shared his mitochondrial DNA with a forty-two-year-old history teacher from the school in nearby Cheddar village. This means that a female relative of the Cheddar Man (say, his sister or his mother, or someone else up the female line from him) was a *direct ancestor* of the present-day local schoolteacher—a direct human lineage of over nine thousand years!

Similar use of mitochondrial DNA combined with the molecular clock has resolved the old debate about whether the hominid known as Neanderthal man was actually a member of the same species as the modern human. Previously, the question was debated by palaeontologists poring over skeletal remains and arguing round and round in circles about differences in cranial capacity and facial structure; but in 1997 the Swedish biologist Svante Pääbo and colleagues at the University of Munich announced that they had managed to obtain DNA from the teeth of Neanderthal fossil remains, and had compared it with that from modern humans. This comparison showed the two types of DNA to be so different as to more or less rule out the possibility that Neanderthals and modern humans might have been interbreeding members of the same species. Since that first paper DNA has been obtained from several more Neanderthal remains and the original conclusion has been confirmed: modern humans and Neanderthals probably shared a common ancestor about six hundred thousand years ago—much further back in human evolution than most experts previously suspected. And it seems that they exchanged no genetic material when they were together in Europe, and are therefore unlikely to have been part of the same species. Whether they even met we still don't know for sure.

The common ancestor between Neanderthal and modern man would have lived in Africa, where fossil evidence suggests that most of the evolution of our species took place. The crucial moment when our ancestors left that continent to colonize the rest of the world has also come under the geneticists' scrutiny: using the molecular clock on both mitochondrial DNA and Y chromosome[4] DNA anthropologists and archaeologists have begun tracing human ancestry across the world. Rebecca Cann[5] (a population geneticist currently at the University of Hawaii) and others have come to the startling conclusion that all humans originating outside Africa can be traced back to the so-called mitochondrial Eve who lived about 150,000 years ago. Comparing the mitochondrial DNA of Africans with that of non-Africans suggested that this "mitochondrial Eve" did indeed live in Africa, where palaeontologists believe humans evolved. As Pääbo himself has written, "all humans are therefore Africans, either residing in Africa or in recent exile."[6]

Going further back in human ancestry, a group at the Washington University School of Medicine in St. Louis have analyzed human chromosome 2 (the second longest) and confirmed what was first suggested over twenty years ago, namely, that this chromosome arose by fusion of two chromosomes that are still separate in the great apes (chimpanzees, gorillas, orangutans). Now, by comparing DNA and using the molecular clock, we can show when these modern groups diverged: the human line of descent diverged from the chimp line about

Gregor Mendel: Planting the Seeds of Genetics

five million years ago, and the two together diverged from macaque monkeys about twenty-three million years ago. People will argue over a million years here or there, but the picture is remarkably consistent. If you want to find the common ancestor between you and your dog, well, we know that it lived just over ninety million years ago. So, however close you feel to Rover, you're not that close. On the other hand, you are a lot nearer to him than you are to a sea squirt (a blob of marine jelly thought to be close to the originators of the vertebrate group). Our probable common ancestor with modern sea squirts lived about five hundred and fifty million years ago, during the early Cambrian period. Just as astronomy is likely to make you feel humble, so too is molecular biology.

Modern genetics is helping scientists plan for the conservation of such unique environments as Mount Kahuzi, in the eastern Democratic Republic of Congo, whose tropical forest is home to one of the last surviving groups of mountain gorillas.

Gene Conservation

Among plants and animals genetic diversity is an essential ingredient in the battle for survival in the wild because it is inherited variation that enables a species to adapt to the pressures imposed by a changing environment. "Adapt or die" is the watchword of evolution. So conservation biologists throughout the

In the 1960s and 1970s, major outbreaks of grassy stunt virus destroyed rice crops and threatened the food supply of millions of people. In recent years scientists used good Mendelian breeding techniques to develop a stunt-resistant variety of cultivated rice (*Oryza sativa*).

world are using techniques of the new genetics to gather information that will enable them to understand and so protect threatened species. In the Albertine Rift of central Africa biologists analyze the DNA of warblers to determine their genetic diversity, and discover, on the way, that what was originally thought to be a single species is actually four genetically distinct groups. Others in the Caribbean plot the familial relationships of sharks in order to understand their mating patterns. It is only with such knowledge in mind that habitats can be preserved in a way that contributes to efficient protection of the species themselves. Zoologist Dr. Kevin Feldheim explains how, by genetically identifying the sharks that they sample, he and his co-workers have found that "females are coming back to specific islands to give birth to their young on a two-year cycle. We also found that females are mating with several males to give rise to their litters. We were only able to answer these questions using DNA technology. So DNA technology has revolutionized how we view biological systems and has challenged some long-standing theories of behavior."

Species conservation is concerned with the whole organism, but it goes hand in hand with gene conservation. Genes are, quite literally, the heritage of the living world, essential in medicine, but also in animal-breeding and crop improvement. Each plant or animal species that becomes extinct carries with it possibly unique genes that we will never see again. As the instructions for making protein lie in the genes, this is the reason why each unique gene is precious and every gene lost to the world is a potential disaster. Currently 15,503 plant and animal species are facing global extinction,[7] most of them through destruction of habitat. Extinction is forever. All the cross-breeding or genetic modification in the world won't bring back lost plants and animals, or the genes that vanished with them. This was what Vavilov understood when he established the first gene bank in Leningrad, in the institute that is now named after him.

There are many cases where a wild plant has been used to confer disease resistance on a crop plant. For example, grassy stunt virus became an important disease of cultivated rice (*Oryza sativa*) in the 1960s and 1970s. Major outbreaks destroyed rice crops in India, Sri Lanka, and Southeast Asia, threatening the food supply of millions of people. The International Rice Research Institute in the Philippines possessed thousands of relatives of rice in its seed bank, so scientists at the Institute began screening these plants for resistance to the virus. After four years' work they came across one species of wild rice, *Oryza nivara*, which was resistant. *Nivara* is of no use for food production itself—in the wild it is a weedlike grass that grows on the edge of marshes in northern India—but it just happens to possess a dominant gene that gives resistance to the virus. Using good Mendelian breeding techniques—hybridizing the resistant rice with the cultivated species, then back-crossing repeatedly with the original cultivated plant—the IRRI was able to develop a stunt-resistant variety of *Oryza sativa* and thus guarantee the food of millions. Mendel would have understood perfectly.

Traditional cross-breeding has a major limitation: you can only move a gene from one variety to another if the two types are sufficiently similar to produce hybrids. A gene in a completely different species will be of no use at all. Modern genetics has the answer. Just as Mendel once grafted one kind of fruit tree onto another, nowadays molecular biologists are able to graft genes from one species to another, completely unrelated, species. Once again, rice provides us with an example. A number of different groups of plants carry genes for what are called "defensins," small proteins that defend the plant against a number of diseases. Examples of such resistance carriers are members of the *Brassica* genus. However, rice plants possess no such genes and are therefore susceptible to a number of diseases, including rice blast (caused by a fungus) and leaf blight (caused by bacteria). Rice blast is particularly serious: every year it destroys enough rice to feed sixty million people. Recently a team from Japan[8] used the techniques of molecular biology to successfully transfer genes for defensin from two species of *Brassica* (*oleracea* and *campestris*) directly into cultivated rice (*Oryza sativa*). The rice plants became resistant to both blast and leaf blight, but apparently suffered no other changes to their genetic makeup. Furthermore they passed the resistance down to their offspring. There is no worry about the safety of the transferred genes: they have been widely tested on human beings for thousands of years: *Brassica oleracea* is the species that give us cabbage, cauliflower, and broccoli, while *Brassica campestris* gives us turnip, oilseed rape, and mustard. Techniques like this are yet another weapon in the armory of those who work to protect and improve food crops—but they all depend on the preservation of plants and their genes, both in the wild and in tissue and gene banks around the world.

Pushing into the Unknown

The Abbey of St. Thomas in Brno is on the other side of Petrov Hill from the main town center. To get there you climb steep, cobbled streets toward the Cathedral of Saints Peter and Paul. The church sits like a black crow at the summit of the hill. Up there cars are few and the noise of the town seems rather remote. You can see for miles across the city that Mendel knew, a city of textile mills in his day, of mechanical and engineering plants now. Ahead of you is the descent down the long street called Pekařská, past the hospital where Mendel himself tended the sick and felt such distress that Abbot Cyrill Napp had to move him to other duties. At the bottom of Pekařská you reach the traffic lights, and there is the abbey in front of you, with the buttresses and steep-pitched roof and Gothic spire of the Church of the Assumption on the right. Little seems to have changed here. The tall windows that face you are those of the abbot's apartment, where Mendel himself once lived. The church is where he celebrated mass on Sundays and where his funeral was held. The hill behind the abbey is where he built his beehive (they still keep bees there) and where, after his death, a bonfire was made of all his precious papers.

To reach the monastery garden you have to walk round to the left of the abbey into the wide square that is named after him, Mendolovo Náměstí, Mendel Square, where the trams clang and grind on their rails. Fifty yards along the high convent wall you reach a gate, and inside is the garden. It is a quiet, contemplative space. There is an expanse of lawn where you can still see the foundations of the greenhouse where he fertilized the peas with such painstaking care. Beyond the lawn is shrubbery, with the memorial that they erected to him in 1910, an idealized statue in white marble, almost like a shrine to a saint. He holds his arms out, not to shelter the poor or the needy, or faithful souls and repentant sinners, but marble pea plants. If you visit in the summer there will be real peas growing in the soil at the base of the statue.

Arabidopsis (commonly known as woody cress) has only recently challenged long-held beliefs about genetic inheritance.

After over a decade of silence in the field of genetics, Mendelism was finally reinstated in Communist Czechoslovakia in the early sixties. In 1965 a small Mendel museum was opened in the building that gives onto the garden, so that the occasional visiting biologist could look round the collection of artifacts that had somehow survived the Stalinist era—Father Gregor's microscope, his telescope for observing sunspots, his meticulous meteorological records. In those early days of thaw the brothers weren't there, of course. They had been expelled by the Communist authorities in 1950 and the building had been handed over to a number of state institutions. But by the middle of the 1960s, when Nirenberg and others in the United States were cracking the genetic code, you could at least mention the name of Mendel in his home country. It wasn't until the fall of the Iron Curtain in 1989 and the return of democratic government that the Abbey of St. Thomas in Brno could finally come back to something like its original role. Restored to its true ownership, the Augustinian friars are now back in residence. A Mendel center has been created, with the permanent exhibition of Mendel's life and work redesigned. There are plans to develop an experimental garden to demonstrate Mendel's discoveries. But apart from the museum, the abbey also supports a series of seminars and lectures on issues connected with genetics and society, and hosts conferences on similar themes. The spirit of the community over which Cyrill Napp and Gregor Mendel once ruled is still alive.

And those genes that Mendel saw vaguely through a blizzard of numbers and ratios and that Morgan perceived specifically as beads on a string? Our picture of them has come a long way. We know now that they are relatively short sequences of DNA bases that together have some kind of meaning for the cell—they can be read to make proteins. They are found at intervals along the length of a DNA molecule—a chromosome—that is comprised of millions of such bases, most of which carry no message at all. From Barbara McClintock's work we know that sometimes bits of DNA jump around the genome, often causing havoc when they do so, and from the discoveries of the 1970s we know that bacteria can chop genes up, copy them, and then hand them round as gifts to other bacteria. We know now that we can enlist these bacterial enzymes to do the same thing ourselves—indeed, these days, such procedures are carried out in undergraduate biology courses. But still the fundamental ideas of Mendelian genetics hold true. Genes are functionally distinct units that are more or less stable and are handed down from parent to offspring in reproduction. They are made of DNA and therefore DNA is the legacy. RNA and the protein that it makes is how an individual spends it, but the legacy is bequeathed by that individual's mother and father in the form of the DNA. That handing down of characteristics was what Gregor Mendel first understood one hundred and fifty years ago. It seemed that this applied to all the living organisms in the world. And then along came *Arabidopsis*.

This small, inconsequential member of the mustard family is to plant genetics what *Drosophila* or the laboratory mouse is to animal genetics. It has a rapid life cycle and has been used in genetics research of all kinds, but it was in March of 2005 that a paper was published in the journal *Nature* that made waves so large that they finally washed *Arabidopsis* up on the front pages of the world's press.[1] A group working at Purdue University in Indiana had discovered that *Arabidopsis* plants homozygous for a recessive mutation called HOTHEAD could, occasionally and totally surprisingly, recreate a normal, functioning dominant version of the gene. If that sounds complicated, it is not: Mendel would have understood the story perfectly well. The notation for the HOTHEAD mutation is **HTH** for the normal allele (gene) and **hth** for the HOTHEAD allele. A plant that is double recessive (**hth/hth**) has flowers in which the male and female parts are rather messily fused. If the plant is heterozygous (**HTH/hth**) then the flowers are normal, because HOTHEAD is recessive. So far, so Mendel. Of course, in the wild most Arabidopsis plants will be **HTH/HTH**.

In 2005 scientists at Purdue University observed some startling genetic behavior in *Arabidopsis*: some plants appeared to inherit genes that their parents did not possess. The team found that the plant was somehow restoring genes from grandparent (or earlier) generations to correct a mutation present in the parent plant. This *Arabidopsis* plant, in comparison to that on the previous page, exhibits the HOTHEAD mutation.

When the Purdue group crossed HOTHEAD mutant plants together (**hth/hth** x **hth/hth**) they expected to get nothing but more mutants. Mendel would have expected exactly the same. It was he who showed, all those years ago, that when plants showing the recessive condition are self-fertilized, they produce only more recessive-type plants. However, in about 10 percent of the offspring from such a cross, the Purdue biologists discovered normal plants—heterozygotes of the **HTH/hth** genotype. It seemed impossible. Where had the normal gene come from? They considered the obvious explanations. Perhaps the **hth** genes had mutated back to the **HTH** form. Perhaps there had been contamination with pollen from other sources (Mendel's old worry). Perhaps the HOTHEAD plants had used a close relative of the **HTH** gene (they have a number of rather similar genes scattered through their chromosomes) to somehow reconstruct a new **HTH**. Even that would have been startling enough, but the Purdue team have managed to discount that possibility and all the other explanations as well. The fact is that they have discovered a phenomenon that they simply can't explain, certainly not within the limits of Mendelian genetics. All they can suggest is that, because all mutant plants have normal plants in their ancestry, somehow the HOTHEAD mutant holds a molecular "memory"—perhaps a molecule of RNA outside the chromosomes—of what the "good" version of the HOTHEAD gene was like and is able to use this to reconstruct a working version of the correct gene. If this is so, in some cases the memory seems to have survived over two or three generations.

The scientific world is still puzzling over this phenomenon. Is it non-Mendelian inheritance, the reconstruction of a gene that appeared to have been lost two or three generations previously? I think Mendel would have watched this particular development in genetics with quiet delight. In 1867 he wrote to Carl von Nägeli:

> I knew that the results I obtained were not easily compatible with our contemporary scientific knowledge, and that under the circumstances publication of one such isolated experiment was doubly dangerous; dangerous for the experimenter and for the cause he represented.

So he understood very well the dangers, and the excitement, of pushing the boat out into unknown waters. And he would surely also have been immensely proud to see that the very first reference in the Purdue team's paper is to one "Mendel, G. *Versuche über Pflanzen Hybriden. Verhandl. Naturforsch. Ver. Brünn* 4, 3–47 (1866)." That a paper written by an amateur scientist in 1866 can still be cited in a major academic paper in 2005 is tribute enough to the father of genetics.

Mendel memorial at the abbey. Although Mendel was overlooked in his lifetime, the friar is celebrated today.

Notes

CHAPTER 1: The Lectures, 1865

1. A secondary school comprising grades 5–10, with an emphasis on practical work and natural science
2. All direct speech in this chapter is taken from the Bateson translation of Mendel's original paper, with adaptations by the present author.
3. Keel: the petals of the pea that enclose the reproductive parts. The stamen makes the pollen; the stigma receives the pollen.
4. Seed plant = female parent; pollen plant = male parent
5. There was one, in the Brünn daily newspaper, *Tagesbote* (Daily Courier).

CHAPTER 2: Childhood

1. This is the date that the family always celebrated and that appears on the inscription, although the parish register records July 20 as his birthday.
2. From 1867 known as the Austro-Hungarian Empire
3. *Gräfin*: countess
4. As a promising pupil Johann was sent, for one year, to the Piarist School at Leipnik (modern Lipník).
5. A higher secondary school
6. Hugo Iltis, *The Life of Mendel*, English translation 1932, second edition 1966
7. Quoted from Mendel's autobiographical sketch written in 1850 to support his application for a teaching certificate. Quaintly, this is written, as quoted, in the third person.
8. *Idem.*
9. From the bill of sale, quoted in Iltis, *op. cit.*
10. See p. 36.
11. Largely quoted from Iltis, *op. cit.*, p. 42. I have substituted "institution" (which the Altbrünn convent is) for "monastery" (which it isn't) as a translation of *Stift*.
12. In German: *fast den Ausgezeichnetsen*

CHAPTER 3: Education

1. Napp had called Diebl in as advisor on modernization of the convent's estates. He was also instrumental in having Diebl appointed to the staff of the Brünn Institute of Philosophy.
2. Fruit-growing and grape cultivation
3. Quoted from Iltis (1966), p. 56
4. Letter of recommendation from the head of the *Gymnasium* at Znaim, May 25, 1850
5. From Mendel's autobiographical note

6. The Doppler effect explains the apparent shift in frequency or wavelength of waves from a moving source, as perceived by a stationary observer. It explains, for example, why the pitch of an ambulance siren changes as it moves nearer or farther from the listener.
7. Apocalypse: The Revelation of Saint John the Divine, New Testament
8. *Hujus*: of this (month)
9. Quoted from Iltis, *The Life of Mendel*

CHAPTER 4: The Scientific Landscape

1. Quoted in Hugo de Vries's doctoral thesis
2. Quoted from Wood and Orel, *Genetic Prehistory in Selective Breeding*, 2001
3. The doubling of the chromosome number, which may make a usually sterile hybrid fertile. In plants it occasionally occurs spontaneously.
4. Napp, 1837, quoted, with adaptation, from R. J. Wood's paper *The Sheep Breeders' View of Heredity (1723–1843)* presented at the Max-Planck Institute for the History of Science Conference, 2003

CHAPTER 5: The Research Program

1. The terms F_1, F_2, etc. are modern and are used here because of their familiarity. They were not used by Mendel, for whom "F_2" would have been the first generation from the hybrid.
2. Mendel's own emphasis in his 1866 paper
3. Darwin's theory of "pangenesis," a great embarrassment to the reputation of the great man. It is found in his book *The Variation of Animals and Plants under Domestication*, 1868. To be fair, it was only a speculative hypothesis without any experimental evidence to support it. For a brief discussion of pangenesis see p. 87.
4. Fisher, 1936. "Has Mendel's Work been Rediscovered?" *Annals of Science* I
5. Darwin, 1868. *The Variation of Animals and Plants under Domestication.* Quoted from the 1888 U.S. edition
6. I have used the translation of Mendel's paper available on the MendelWeb at: http://www.mendelweb.org/MWpaptoc.html. The reader is encouraged to visit this valuable site for further information.
7. *inneren Beschaffenheit* in the original
8. Spermatophytes. Mendel actually used the now obsolete term "Phanerogram."
9. *der Elemente* in the original

CHAPTER 6: The Follow-up

1. According to Orel in his 1996 biography of Mendel, *Gregor Mendel, the First Geneticist*. The information about the fate of the offprints is taken directly from this book.
2. *Die Pflanzenmischlinge*
3. Quoted from Fisher, 1932, "Has Mendel's Work Been Rediscovered?" *Annals of Science* I
4. Quoted from Iltis
5. Quoted by Mendel himself in his reply to von Nägeli
6. Stock. *Zea* is the genus that includes maize or corn. *Mirabilis* is a South American plant known as Four-o'clock. *Lychnis* are campions (synonym: *Silene*).
7. This author's emphasis
8. Letter to von Nägeli, July 3, 1870
9. Letter to von Nägeli, September 27, 1870
10. Letter to von Nägeli, November 18, 1873

CHAPTER 7: Life beyond Peas

1. The details come from a letter from Mendel to Leopold Schindler, husband of his sister Theresia.
2. The paper published in 1870 giving an account of his attempts to hybridize *Hieracium* is of no more than curiosity value.

CHAPTER 8: The Rediscovery

1. Fleeming Jenkin, June 1867. (Review of) "The Origin of Species," *The North British Review*, 46
2. Jean Baptiste Lamarck, a naturalist from northern France, began propounding his theories of inheritance of acquired traits in 1801.
3. Kossel received a Nobel Prize for this work in 1910.
4. In the original, *la loi de disjonction des hybrides*
5. Robert Olby, *Origins of Mendelism*, 1966
6. English translation of the paper "*G. Mendels Regel über da Verhalten der Nachkommenshaft der Rassenbastarde*" from Stern & Sherwood, 1966. Correns's own italics
7. Correns's own emphasis

CHAPTER 9: The Mendel Legacy

1. Morgan, T. H. 1910. "Sex-limited inheritance in *Drosophila*," *Science* 32: 120–22.
2. The notation is that used by Sturtevant (see pp. 108–9)—both rudimentary and vermilion are recessives.
3. Sturtevant, A. H. , 1913. " The Linear Arrangement of Six Sex-linked Factors in *Drosophila*, as Shown by Their Mode of Association," *Journal of Experimental Zoology*, 14: 43–59

4. In his paper, Sturtevant makes an elementary error in addition, which gives an incorrect value of 26:9 percent between **P** and **M**. This is repeated throughout the paper and results in an inaccurate map (57.6–30.7 = 26.9). Youthful carelessness?
5. Garrod, quoted from the end of the first chapter of *Inborn Errors of Metabolism*, 2nd edition
6. Muller, H. J. 1922. "Variation Due to Change in the Individual Genes," *The American Naturalist*, 56: 32–50
7. Levene and Bass, *Nucleic Acids*, 1931
8. Chargaff, *Annals of the New York Academy of Sciences*, 1979
9. From Wilson's book, *The Cell*, 1925

CHAPTER 10: New Currents in Genetics

1. Published in 1872 in the *Fortnightly Review*. Galton came to the conclusion that prayer doesn't work. Part of the evidence? Sovereigns, who are regularly prayed for by millions of their subjects (the British version of the Anglican Book of Common Prayer includes prayers for the British monarch), die younger than their upper-class subjects!
2. The eugenicists always seemed to regard their own personal attributes as the most desirable of all, although, ironically, Galton and his wife never had children, so the great man's talents were never passed on to a second generation.
3. The post is currently held by the well-known popular writer on genetics, Steve Jones.
4. The Cold Spring Harbor Laboratory is still one of the world's leading genetics research laboratories, with its eugenics past consigned to history. James D. Watson (director 1968–1994) is currently its chancellor.
5. Now known as *Streptococcus pneumoniae*
6. **S** (smooth) and **R** (rough) refer to the appearance of the bacterial colonies when grown on an agar plate. There are also four different types of the bacterium, I, II, III and IV. Transformations were also achieved among these different types.
7. Avery's emphasis
8. Avery, MacLeod, and McCarty, 1944. "Studies on the Chemical Nature of the Substance Inducing Transformation of Pneumococcal Types," *Journal of Experimental Medicine*, February 1944
9. Dated January 20, 1945
10. Howard A. Schneider, quoted by Maclyn McCarty in his book, *The Transforming Principle*
11. Joshua Lederberg again, on the thirty-fifth anniversary of the publication of the paper
12. X-ray crystallography, devised 1912–14 by William

Bragg and his son Lawrence, involves firing a beam of X-rays into a crystal and recording how the rays are scattered (diffracted). From the pattern that emerges clues can be obtained about the arrangement of atoms in the molecules that make up the crystal.

13. From Chargaff's paper of 1950: Chemical specificity of nucleic acids and mechanism of their enzymatic degradation. *Experientia* 6: 201–9

14. It had *three* phosphate-sugar strands, rather than the two of the final model, and the phosphate groups were on the inside. Franklin always insisted that they had to be on the outside in order to stabilize the molecule.

15. Since May 1952, in fact. The account here follows Wilkins's published version in his biography.

16. Letter to *Nature*, 171 (1953), 737–38

CHAPTER 11: From Code to Genome

1. Bacteria don't actually possess nuclei. Their DNA, in a continuous loop rather than being divided into distinct chromosomes, lies in their cytoplasm.

2. In RNA the base uracil (U) takes the place of thymine (T). So RNA base-pairing is G: C and A: U.

3. Along with his twin brother, Roy, a historian, Medvedev was one of the leading Soviet dissidents of the postwar period.

4. From the Nirenberg papers on deposit at the National Library of Medicine, at the National Institutes of Health at Bethesda, Maryland

5. Nirenberg shared the 1968 Nobel Prize with Gobind Khorana, along with Richard Holley, who had been the first to isolate and sequence a transfer RNA molecule—Crick's proposed "adaptor molecule." One can't help feeling that a single prize was being spread rather thin that year!

6. The genome of any organism is its entire length of DNA, including all the genes and, in the case of higher organisms, all the extra DNA that doesn't actually form genes (see p. 152).

7. Being a virus, phi-X 174 doesn't qualify as an organism. Its DNA is 5375 bases long.

8. Better termed "non-coding" DNA

CHAPTER 12: Where Do We Go from Here?

1. Developed by Alec Jefferies of Leicester University in 1984

2. 1707–78. Known to the world as Linnaeus, von Linné was the father of biological classification.

3. Mitochondria are the cell organelles that carry out aerobic respiration.

4. The Y chromosome is the one that determines maleness, and is passed down the male line in any family tree rather like the way mitochondrial DNA is passed down the female line. It was Y chromosome data that showed that Thomas Jefferson almost certainly fathered a son (Easton Hemmings) by his slave Sally Hemings, and probably some of her other children as well.

5. "Mitochondrial DNA and Human Evolution," *Nature* 325 (1987), 31–36

6. *Science*, vol. 291, no. 5507, 1219–20, February 16, 2001

7. World Conservation Union (IUCN) Red List for 2004

8. Kawata et al., *Japan Agricultural Research Quarterly*, vol. 37, 2003

CHAPTER 13: Pushing into the Unknown

1. Lolle et al., "Genome-wide Non-Mendelian Inheritance of Extra-genomic Information in *Arabidopsis*," *Nature*, 2005

Acknowledgments

My journey into the world of Mendel started over a decade ago in the city of Brno and I must in some sense thank the city as a whole, as well as my wife, who first suggested that we travel there. She is unstinting in her support of my writing and that fortuitous visit engendered a novel, *Mendel's Dwarf*, and, indirectly, this work of nonfiction.

Individual thanks for this book must go to the staff of the Mendel Museum, in particular Anna Nasmyth, who has done so much to put the museum on the map. While in Brno, and later, I also benefited from the advice and help of that doyen of Mendel experts, Vítězslav Orel. He has been generous with his time, and apart from personal contact, his definitive life of Gregor Mendel, *Gregor Mendel, The First Geneticist*, has been invaluable. Vítězslav's colleague, Roger Wood of Manchester University, made a number of useful papers available to me and their jointly authored book *Genetic Prehistory in Selective Breeding* provided invaluable information on the state of breeding in the period prior to Mendel's work.

The Field Museum in Chicago was instrumental in setting up the book in the first place and I have only ever received enthusiastic help from Museum staff, both in Brno and in Chicago itself. In particular I must thank Cheryl Bardoe who planted the first seed, and Tiffany Plate who, along with Cheryl, has followed the book to its final blooming. Mark Alvey always provided useful ideas and a dry sense of humor that has kept me amused; and Shannon Hackett made many astute and invaluable comments. In Rome, Toby Hodgkin of the International Plant Genetic Resources Institute pointed me in the direction of useful examples of modern gene conservation.

A number of people have read the book in part or in whole, and have offered recommendations and advice. Elof Carlson of the State University of New York made suggestions of particular value. I must also thank Eric Himmel of Harry N. Abrams, Inc., who brought an expert but non-Mendelian viewpoint to the book when it was in draft form.

Occasionally I may not have acted on advice given, but that is the prerogative of an author. With that prerogative goes responsibility: any opinions, unless they are specifically attributed to others, and all errors, are mine alone.

Simon Mawer

The Field Museum is delighted to co-publish *Gregor Mendel: Planting the Seeds of Genetics* with Harry N. Abrams, Inc., and to bring to light the fascinating life of Gregor Mendel and his important contributions to heredity that laid the groundwork for the field of genetics.

The stories of Gregor Mendel and his successors show us that science is a truly human endeavor. From Mendel's painstaking work in an Abbey garden in the 1850s to scientists analyzing samples in The Field Museum's Pritzker DNA Laboratory today, our knowledge of genetics is the cumulative result of many individuals' insight, industry, and passion to learn more about the world.

As exciting as scientific discoveries are, it is also humbling and inspiring to realize that the more we learn about our world, the more there is still to be investigated. We hope that projects like this book, and the corresponding exhibition, will help strengthen scientific education in our elementary and secondary schools. Many young people are being "turned off" from the study of science, mathematics, and engineering. Scientific literacy and public discussion provide the foundation for national competitive success in an increasingly "flat world"—to use Tom Friedman's phrase. If we are helpful to teachers of science in the classroom and to family discussions around the dinner table, we will be fulfilling our purpose—"to explore the earth and its peoples."

This book would not have been possible without the help of many individuals. Our thanks go to author Simon Mawer, who has produced a brilliant manuscript on Mendel and the course of genetics in the twentieth century. Field Museum Exhibition Director Robin Groesbeck guided the project from its inception. Project coordinator Tiffany Plate, with exhibition project manager Cheryl Bardoe, organized the book's reviews, wrote captions, and supervised the selection and acquisition of photographs for the book. Members of the exhibition team, including Mark Alvey and Tim Mikulski, provided key assistance for this volume. Photographs and illustrations were provided by Greg Mercer, Mary Goljenboom, Rachel Post, Jerice Barrios, Nina Cummings, Michael Godow, and John Weinstein.

Field Museum research staff contributing to the development of this book include Dr. Shannon Hackett, Associate Curator and Head, Division of Birds, and Dr. Kevin Feldheim, Manager, Pritzker Laboratory. We also wish to acknowledge the following individuals who provided content reviews: Dr. Elof Carlson, Dr. Michael Dietrich, Dr. Rochelle Esposito, Vítězslav Orel, and Dr. Iris Sandler.

Finally, this project would not have been possible without the participation of Abbot Lukas Evzen Martinec, of the Abbey of St. Thomas, Brno and Anna Nasmyth, Project Manager for the Vereinigung zur Förderung der Genomforschung.

John W. McCarter, Jr.
President and CEO, The Field Museum

Index

Numbers in *italics* refer to illustrations.

Illustration Credits

(Images from The Field Museum are credited as "FM.")

Courtesy of the American Philosophical Society: 110, 111 (both), 113, 114, 122, 137, 148, 149; Courtesy of Archief Bibliotheek Biologisch Centrum, University of Amsterdam: 98; Courtesy of the John Innes Centre Archives, William Bateson Collection: 102; Peter Bernhardt, Plant Cell Biology Research Center, University of Melbourne, Australia: 79 (right); Boveri Archive, Würzburg University: 94–95; © A. Barrington Brown / Photo Researchers, Inc: 138; Tom Campbell, Purdue University: 163; Cold Spring Harbor Laboratory Archives: 135; Used with permission of the University Archives and Columbiana Library, Columbia University: 108; Portrait from *Leben und Wirken von Carl Wilhelm von Nägeli* (Life and Work of Carl Wilhelm von Nägeli) by C. Cramer, 1896: 71 (bottom); Dibner Library of the History of Science and Technology, Smithsonian Institution Libraries, Washington, D.C.: 35; © FM/Robin Foster: 158; Field Museum Division of Insects: 38; © FM/ Mark Alvey: 18; © FM/GN90808; From *The Botanical Magazine or Flower-Garden Displayed:* 8 (from Vol. 2), 77 (from Vol. 11); © FM/GN90795d; From *Contributions to the Genetics of Drosophila Melanogaster:* 104; © FM/GN90803d; From *English Botany; or Coloured Figures of British Plants,* Third Edition, Volume I: 76; © FM; From *Erinnerungen an Johann Gregor Mendel, mit einem Faksimile des Manuskriptes von Mendels Arbeit Versuche uber Pflanzenhybriden:* 70; © FM/ GN90800d; From Fremenville: 82; FM/GN90794d; From *Ladies' Flower-Garden:* 28; From *Flora von Deutschland, Oesterreich, und der Schweiz* Volume Dritter Band (Volume 3): © FM/GN90791d: 68; © FM/GN90792d: 90; © FM/ GN90793d: 52; ©FM/GN90801d; From Ledermuller's *Mikroskopische Beobachtungen,* Band II: 14; © FM/GN90805d; From *Oeuvres Completes de Buffon Planches,* Tome I: 47; © FM/ GN90804d; From Joseph Paxton's *Magazine of Botany and Register of Flowering Plants,* Volume 11: 64; © FM/GN90790d; From Mendel's *Principles of Heredity:* 107; © FM/GN90796d: From *The Third Chromosome Group of Mutant Characters of Drosophila Melanogaster:* 109; © FM/ GN90834d: From *Transactions of the Horticultural Society of London* Volume I, 2nd Edition, 1835: 16; © FM/Greg Mercer: 13, 54 (both), 57, 61, 144, 153; © FM/John Bates: 157; © FM/Tim Mikulski, based on a work from Wellcome Trust Medical Photographic Library, at www.wellcomw.ac.uk/en/genome/the genome/ hg02b001.html: 147; © FM/John Weinstein/Z94281_8c: 154; © FM/John Weinstein/GN90811.91d: 140; © FM/John Weinstein/GN90811_48d; 152; Photo montage © FM (by photographer left to right, descending): William C. Burger, John Bates, William C. Burger, Sabine Huhndorf; Paul Heideman, Robert Luecking, William C. Burger, Robert Luecking; Robert Luecking, William C. Burger, Robert Luecking, Peter Batson; John Weinstein/ZN981929c, William C. Burger, Robert Luecking, Robert Luecking; William C. Burger, Mark Westneat, William C. Burger, Robert Luecking: 150; W. Flemming, *Zellsubstanz, Kern und Zellheilung,* 1882: 94 (bottom); "Diagram of Heredity" [Letters to the Editor] by Francis Galton. *Nature,* vol. 57, #1474, Jan. 27, 1898, p. 293: 121 (top); From Nicolaas Hartsoeker's *Essai de dioptrique,* published in Paris, 1694: 15; Courtesy Eric Himmel: 59; Courtesy University of Illinois Archives, Iltis Mendeliana Collection: 26; © IMAGNO/Austrian Archives: 19, 36–37; © IMAGNO/Ullstein: 23; Portrait from Report of the Third International Conference on Genetics, 1906, W. Wilks, editor: 97; From *The Kallikak family; A Study in theHeredity of Feeblemindedness,* by Henry Herbert Goddard, 1912: 123; Courtesy Mendel Museum: 165; Mendel Museum photo by Stepan Bartos: title spread (both images); 4, 11 (top), 21, 22, 25, 30, 31, 40, 41, 42, 45, 46, 50, 55, 56, 71 (top); 73, 74, 78, 79 (left), 84 (original painting by Alois Zenker, 1884), 85, 86–87, 88, 89, 92; Courtesy of the Mendelianum, Moravian Museum, Brno, Czech Republic: 10, 11 (bottom), 12, 33, 87; Courtesy of the National Library of Medicine: 44 (right), 48, 96, 100, 117, 120, 125, 134 (top), 143; Courtesy Vítězslav Orel: 32; From the Ava Helen and Linus Pauling Papers, Special Collections, Oregon State University: 118, 136; Courtesy of the Rockefeller Archive Center: 116; Reproduced from *The Journal of Experimental Medicine,* 1944, vol. 79, 158 (figure 1) by copyright permission of The Rockefeller University Press. Photo by Mr. Joseph B. Haulenbeek, courtesy of The Oswald T. Avery Collection, National Library of Medicine: 127; © Rothamsted Research: 49; Reproduced with the kind permission of the Director and the Board of Trustees, Royal Botanic Gardens, Kew: 6; Royal Horticultural Society, Lindley Library: 51; © The Royal Society: 134 (bottom); Portrait from *Studien, Populäre vorträge von M. J. Schleiden,* 1857: 44 (left); From *Socialist Agriculture,* May 21, 1937, no. 114 (2502):3. Photo courtesy of Valery N. Soyfer: 128; Figures from "On the Morphology of the Chromosome Group in *Brachystola magna*" by W. S. Sutton. *Biological Bulletin* vol. 4, no. 1, Dec. 1902: 106; Dan Tenaglia - missouriplants.com: 160; Manuscripts/Rare Book Division, Universitätsbibliothek Tübingen: 93; Courtesy of N.I. Vavilov Research Institute of Plant Industry: 130; Courtesy Wellcome Trust Medical Photographic Library: 121 (bottom); From *Zeitschrift für Ärztliche Fortbildung,* 33 (1936): 115. Reproduced in *Racial Hygiene: Medicine Under the Nazis* by Robert Proctor, 1988, Figure 29: 129; Marcia Ponce de León and Christoph Zollikofer, University of Zürich, Switzerland: 155.